I0055474

Analog Automatic Control Loops in Radar and EW

The Artech House Communication and
Electronic Defense Library

Arbenz, Kurt, and Jean-Claude Martin, **Mathematical Methods of Information Transmission**

Arbenz, Kurt, and Alfred Wohlhauser, **Advanced Mathematics for Practicing Engineers**

Deavours, Cipher A., and Louis Kruh, **Machine Cryptography and Modern Cryptanalysis**

Deavours, Cipher A., et al., eds., **Cryptology Yesterday, Today, and Tomorrow**

Fielding, John E., and Gary D. Reynolds, **RGCALC: Radar Range Detection Software and User's Manual**

Rubin, Olis, **Analysis and Synthesis of Logic Systems**

Schleher, D. Curtis, **Introduction to Electronic Warfare**

Torrieri, Don J., **Principles of Secure Communication Systems**

Wiggert, Djimitri, **Codes for Error Control and Synchronization**

Wiley, Richard G., **Electronic Intelligence: The Analysis of Radar Signals**

Wiley, Richard G., **Electronic Intelligence: The Interception of Radar Signals**

Wiley, Richard G., and Michael B. Szymanski, **Pulse Train Analysis Using Personal Computers**

Analog Automatic Control Loops in Radar and EW

Richard Smith Hughes

Artech House

Copyright © 1988

ARTECH HOUSE, INC.
685 Canton Street
Norwood, MA 02062

All rights reserved. Printed and bound in the United States of America. No part of this book may be reproduced or utilized in any form or by any means, electronic or mechanical, including photocopying, recording, or by any information storage and retrieval system, without permission in writing from the publisher.

International Standard Book Number: 0-89006-321-4
Library of Congress Catalog Card Number: 88-71763

10 9 8 7 6 5 4 3 2 1

JANET,
MY WIFE AND MENTOR

CONTENTS

Preface ix
Chapter 1 - Automatic Gain Control 1
  Introduction 1
  Dynamic Regulation 8
  Loop Rise Time and Bandwidth (f_{3dB}) 14
  Automatic Gain Control Design Verification 16
  Square Law Detector Test Circuit 16
  Linear Detector Test Circuit 28
  Variable Gain Elements 30
  P-N Junction Diodes as Variable Gain Elements 40
  The Four Basic P-N Diode AGC Configurations 45
  PIN Diodes as Variable Gain elements 49
  GaAs FET as Variable Gain Elements 55
  Applications 58
  Conical Scanning AGC Loop 64
  Optimizing Loop Rise Times 74
  Variable Time Constant 75
  Linearized Time Constant 75
  Basic Loop Stability 80
  References 82
  Appendices:
    1A. Power-Voltage Relationships for a 50-OHM System . 83
    1B. Basic Detector Characteristics 84
    1C. Static Regulation Calculations 93
    1D. AGC Gain Calculation 96
    1E. Loop Rise Time Calculation 99
  Nomenclature 104
  Bibliography 109

Chapter 2 - Automatic Noise Tracking Loops 111

   References 134

   Appendices:

     2A. The Effect of Receiver Gain on Signal Sensitivity

       and Noise 135

     2B. Derivations of $T_{SS}\big|^{GR}_{dBm}$, $T_{SS}\big|^{Max}_{dBm}$, and

       $G_R\big|^{Max}_{dB}$ Equations 163

     2C. Derivations of $T_{SS}|_{dBm}$, $T_{SS}\big|^{Max}_{dBm}$, and

       $G_R\big|^{Max}_{dB}$ for $B_V \le B_R \le 2B_V$ 169

   Nomenclature 171

   Bibliography 173

Chapter 3 - Range-Tracking Loops 175

   Second-Order, Type II, Range-Tracker Design Procedure . . 185

   Second-Order, Type I, Ranging Tracking Loop 213

   Type I Tracking Loop Design Equations 215

   References 223

   Appendices:

     3A. Range-Tracking Loop and Phase-Locked Loop . . . 224

       Analogy

     3B. Range Discriminator Analysis 229

   Nomenclature 238

Index 241

PREFACE

Automatic gain control, signal thresholding, and range tracking are basic building blocks of many radar and electronic warfare (EW) systems, active and semi-active missiles, and antiradiation missiles (ARMs). This book is meant as a practical starting point for those who must analyze and design these blocks. The reader may well ask why analog, and not digital, automatic tracking loops? Certainly most modern radar, EW, and related systems employ digital (software) technology to perform many of the functions once performed by analog circuitry; however, the digitizing of analog functions requires a fundamental knowledge of analog circuit design and operation. The intent of this book is to help in providing that knowledge. The deviation from pure analog circuitry to a hybrid of analog, digital, and software is up to the readers' requirements; however, it is hoped that this book will provide a solid foundation upon which one can begin to build.

Chapter 1 presents automatic gain control from a theoretical and practical standpoint. Static regulation, dynamic regulation, loop bandwidth, and rise time are covered and the theoretical results verified with several design examples. Methods of varying IF and RF gains are discussed, as are the effects of square law and linear detection on the various AGC parameters and the effect of self-AGC (the IF or RF signal having a controlling influence on the circuits' gain).

Virtually all radar, EW, and related systems employ signal thresholding to identify the presence of a signal. Signal thresholds can be as simple as a comparator and a fixed signal threshold voltage or as complicated as circuitry that nulls any returns due to clutter or noise. Automatic thresholding loops to remove clutter are usually quite complicated and indeed a digital approach at the offset of design is often necessitated. Chapter 2 concentrates on the necessity of providing a noise riding threshold to compensate for the excess noise due to IF/RF amplification preceding a basic crystal video receiver. A fairly simple and straightforward noise tracking threshold is presented to illustrate the techniques involved and the results that may be expected. An automatic noise tracking loop to optimize the detection process is often a necessity, even for simple receivers. This chapter is not meant as an all-inclusive presentation on signal thresholding, but rather a starting point for those who must define and design this important building block.

No book with the title *Analog Automatic Tracking Loops* would be complete without a presentation of range-tracking loops. Chapter 3 presents the analysis and design of both Type I and Type II range-tracking loops from a phase-locked loop perspective. Examples are presented to verify the design theory given.

Topics that deal with signal acquisition and reacquisition, signal duty cycle dependence, etc., have been purposely left untouched. As an engineer employed by the U.S. Government, I am acutely aware of the possibilities of inadvertently and unwittingly compromising Navy programs. To remove this risk, any topic which I felt could, in any way, compromise the work in which we are engaged at the Naval Weapons Center (NWC) was left out completely.

The author would like to extend formally his appreciation to John Daugherty and Paul Hilliard, who during the past 18 years, had the responsibilities of constructing and testing the circuits presented. Ms. Janet Pande, Bob Sutton, and Brad Wiitala reviewed this work, and I am most grateful for the time spent and suggestions offered. I wish to give special thanks to Mrs. Freddie Perry, my NWC editor, who had to take my unintelligible notes and transform them into a readable document.

This work is an evolution of design notes, NWC publications, and other documents, spread over the past twenty years. I wish to express my deepest appreciation to the management of NWC (past and present) who continue, with ever increasing money constraints, to create the practical and academic atmosphere that make works such as this possible.

A final thank you goes to the reader. Without your support, books such as this would not be possible. I hope this work provides some of the material you have been seeking. If it does, the past six months of nights and weekends will have been worthwhile.

Richard Smith Hughes
June 1988

Chapter 1

AUTOMATIC GAIN CONTROL

Introduction

Many radars and electronic warfare (EW) systems employ automatic gain control (AGC) to normalize the received signal prior to signal processing (i.e., range tracking, signal acquisition, direction finding, etc.). This chapter presents the basic AGC theory and design philosophy from a practical standpoint. The theory covers static and dynamic regulation and AGC rise time, with respect to square law detection and linear detection. Various methods of variable gain control are presented and self-AGC (the controlled signal level effecting the gain) is discussed. The chapter concludes with several design examples.

Figure 1-1 illustrates a basic radar receiver employing AGC. As the received input signal varies, the input to the intermediate frequency (IF) amplifier changes. The AGC loop notes the change and varies the gain of the IF amplifier in such a way that the output of the detector remains constant (the AGC voltage could also vary the gain of the radio frequency (RF) amplifier).

Figure 1-2 illustrates the basic components of AGC tracking loops (continuous wave (CW) inputs will be assumed for now; however, operation with pulse inputs is basically the same and will be covered later). The only difference between Figures 1-2a and 1-2b is that one uses a low-pass filter (LPF) and the other an integrator. The differences between these two techniques are also discussed.

1

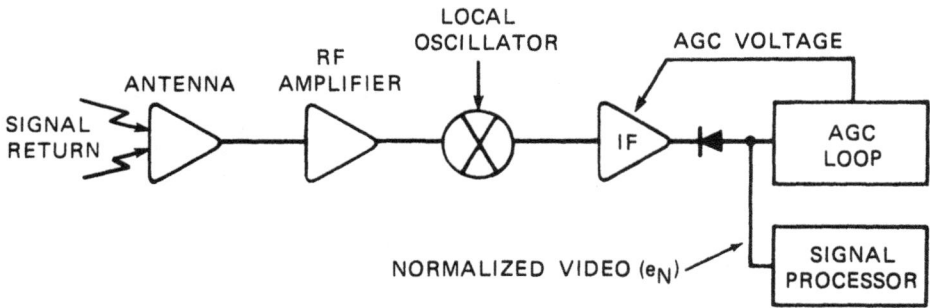

FIGURE 1-1. Basic Radar Receiver.

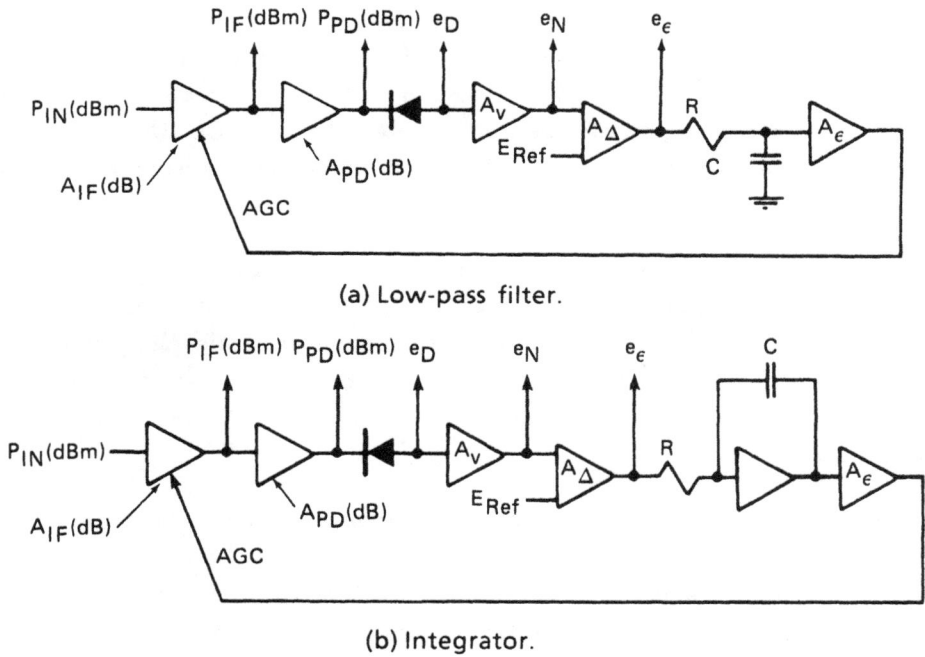

(a) Low-pass filter.

(b) Integrator.

FIGURE 1-2. Basic Components of Automatic Gain Control
(AGC) Tracking Loops.

2

The configurations of Figure 1-2 are quite general in that most AGC loops can be reduced to the components illustrated (necessary pulse stretching and timing circuitry for pulse AGC operation are not shown in the interest of simplicity). The input signal, P_{in}(dBm), is amplified by the variable gain IF amplifier. The variable gain IF output may be amplified or attenuated, depending on the desired normalized detector input power, $P_{PD,N}$(dBm). The detected signal is amplified by the video amplifier, A_v, compared with a reference voltage, E_{Ref}, and again amplified by the error amplifier, A_ε. The resultant voltage drives the variable gain IF AGC input.

If P_{in}(dBm) should increase, P_{PD}(dBm), and thus the normalized video voltage, e_N, would increase, increasing the AGC voltage, and thus decreasing the gain until $e_N = E_{Ref}$.

Three basic parameters define the operation of an AGC loop:

1. Static regulation is the capability to compress large input variations into small output variations. This is the same concept as line regulation in a regulated power supply. The compressed output variation, ΔP_{PD}(dB), divided by the input variation, ΔP_{in}(dB), is called the compression ratio (CR). The change in the normalized video voltage, Δe_N, depends on detector type and will be discussed as we proceed.

$$CR_{IF} = \frac{\Delta P_{PD}(dB)}{\Delta P_{in}(dB)} \qquad (1\text{-}1)$$

$$CR_{vid} = \frac{\Delta e_N(dB)}{\Delta P_{in}(dB)} \qquad (1\text{-}2)$$

2. Dynamic regulation is the capability of an AGC loop to reduce the dynamic input modulation, MP_{in}, appearing at the output, MP_{IF}, and is called the input modulation reduction (IMR). This quality is especially important in conical scanning radars. IMR is dependent on the loop gain (LG) of the AGC loop and may be given as

$$IMR = \frac{MP_{IF}}{MP_{in}} \simeq \frac{1}{1+LG} \qquad (1\text{-}3)$$

3. Loop rise time (τ_r) is the 10 to 90% loop-response time resulting from a step change in input power.

These three parameters are discussed in the next sections.

Static Regulation

Assume that the variable gain IF amplifier has a variable gain characteristic as illustrated in Figure 1-3. (Most variable gain IF/RF amplifiers have a rather linear relationship between gain (dB) and AGC voltage.) The equation relating IF gain to AGC voltage is

$$A_{IF}(dB) = A_{o}(dB) - X(AGC\ Voltage) \qquad (1\text{-}4)$$

where

$$X = \text{variable gain slope in dB/V}$$
$$A_{o}(dB) = \text{maximum gain}$$

AGC loops involve power levels and ratios in the IF and detector portion and voltage levels and ratios in the video (or post detection) portion. Appendix 1A summarizes power-voltage relationships for easy reference.

One primary function of an AGC loop is to keep the output power, P_{IF}(dBm) (or video voltage, e_N), normalized to within a specified amount, ΔP_{IF}(dB) or Δe_N(dB), despite large variations in the input power, ΔP_{in}(dB). A predetector amplifier may be necessary to ensure that the detector is operated at the desired level, linear or square law (see Appendix 1B for detector characteristics pertinent to AGC design).

4

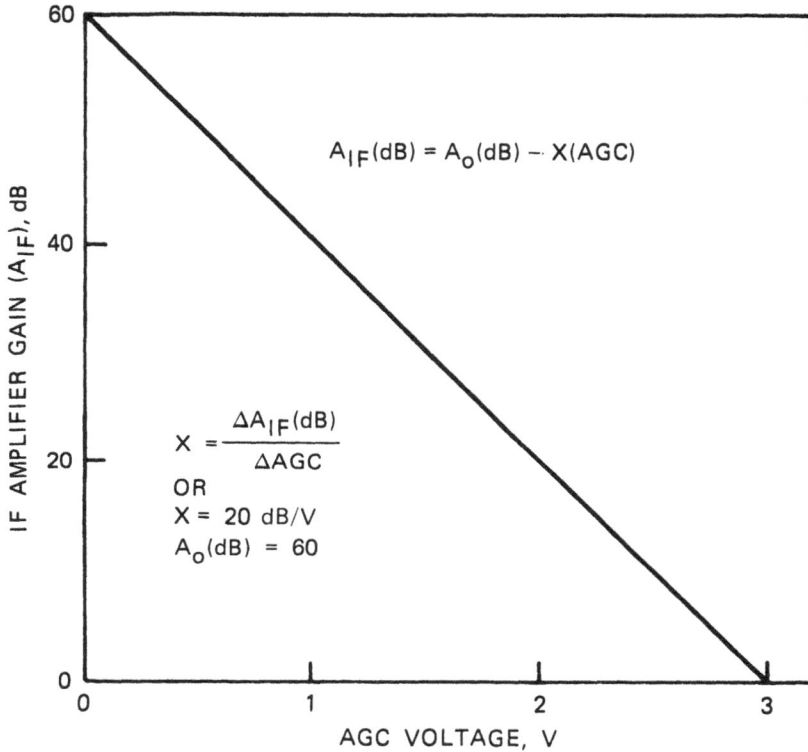

FIGURE 1-3. IF Amplifier Gain Versus AGC Voltage.

The static regulation characteristics of the low-pass filter AGC (Figure 1-2a) will now be discussed. Compression ratio has been defined in Equation (1-1). Appendix 1C presents the solution for ΔP_{IF}(dB) as a function of detector type, linear (Lin) or square law (SL), and the results are

$$\Delta P_{IF,SL}(dB) = 10 \log \left| \frac{\Delta P_{in}(dB)}{X A_\Delta A_\varepsilon e_N} + 1 \right| \qquad (1\text{-}5)$$

$$\Delta P_{IF,Lin}(dB) = 20 \log \left| \frac{\Delta P_{in}(dB)}{X A_\Delta A_\epsilon e_N} + 1 \right| \qquad (1\text{-}6)$$

Thus, to minimize the change in output power, e_N, A_Δ, and A_ϵ must be made as large as practicable (X is assumed constant for a given IF amplifier).

Solving Equations (1-5) and (1-6) for $A_\Delta A_\epsilon e_N$,

$$(A_\Delta A_\epsilon e_N)_{SL} = \frac{\Delta P_{in}(dB)}{X \left(10^{\frac{\Delta P_{IF}(dB)}{10}} - 1 \right)} \qquad (1\text{-}7)$$

$$(A_\Delta A_\epsilon e_N)_{Lin} = \frac{\Delta P_{in}(dB)}{X \left(10^{\frac{\Delta P_{IF}(dB)}{20}} - 1 \right)} \qquad (1\text{-}8)$$

Thus, since X and $\Delta P_{IF}(dB)$ are known, the necessary $A_\Delta A_\epsilon e_N$ may be found.

To illustrate the discussion thus far, consider the following:

Minimum input for AGC action (or AGC delay) $P_{in,min}(dBm) = -70 \text{ dBm}$

Minimum output under AGC action $P_{IF,min}(dBm) = -0.5 \text{ dBm}$

Maximum output under AGC action $P_{IF,max}(dBm) = +0.5 \text{ dBm}$

Maximum input for AGC action (or AGC dropout level) $P_{in,max}(dBm) = -20 \text{ dBm}$

The input dynamic range is

$$\Delta P_{in}(dB) = P_{in,max}(dBm) - P_{in,min}(dBm) = 50 \text{ dB} \qquad (1\text{-}9)$$

and the output dynamic range is

$$\Delta P_o(dB) = P_{IF,max}(dBm) - P_{IF,min}(dBm) = 1 \text{ dB} \qquad (1\text{-}10)$$

Thus the compression ratio is

$$CR = \frac{\Delta P_{IF}(dB)}{\Delta P_{in}(dB)} = 0.02 \qquad (1\text{-}11)$$

or the output increases 0.02 dBm for each 1 dBm increase to the input.

Figure 1-4 illustrates the characteristics of the AGC system just presented.

This discussion has concerned the AGC of Figure 1-2a. Figure 1-2b illustrates an AGC loop that incorporates a true integrator. A true integrator has a very large gain at low frequencies; therefore, under normalized, unmodulated inputs, $e_{\varepsilon} = 0$. Thus,

$$\Delta P_{IF}(dB) \simeq 0 \qquad (1\text{-}12)$$

and perfect regulation is obtained. (Further in the chapter the theoretical equations presented will be verified with a practical example.)

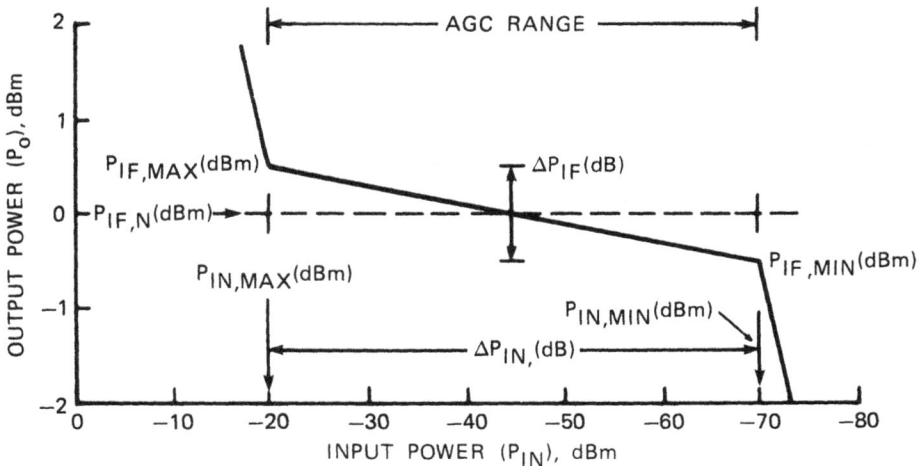

FIGURE 1-4. Typical AGC Characteristics.

7

The next section presents the behavior of the AGC loop, illustrated in Figure 1-2, under input modulation conditions.

Dynamic Regulation

Figure 1-5 illustrates the AGC loop of Figure 1-2a in classical feedback form. Using conventional feedback theory, Oliver [1]* has shown that

$$\frac{\Delta e_{IF}(PP)}{e_{IF}(PP)} = \frac{1}{1 + AB}\left(\frac{\Delta e_{in}(PP)}{e_{in}(PP)}\right) \tag{1-13}$$

where

$e_{IF}(PP) = \text{peak} - \text{to} - \text{peak IF output voltage}$

$\dfrac{\Delta e_{IF}(PP)}{e_{IF}(PP)}\,(100) = \text{percent output modulation } (M_o \text{ or } M_{IF})$

$\dfrac{\Delta e_{in}(PP)}{e_{in}(PP)}\,(100) = \text{percent input modulation } (M_I)$

$AB = \text{loop gain (LG)}$

Equation (1-3) may be written as (assuming LG $\gg$ 1)

$$M_{IF} = \frac{M_I}{LG} \tag{1-14}$$

or, the IF output modulation is reduced by the inverse of the loop gain. Thus the input modulation reduction is (for large loop gains)

---

*Numbers in brackets [ ] refer to references. These references, along with a bibliography, are contained at the end of each chapter.

8

$$IMR \simeq \frac{1}{LG} \qquad (1\text{-}15)$$

and

$$M_{IF} = IMR\,(M_I) \qquad (1\text{-}16)$$

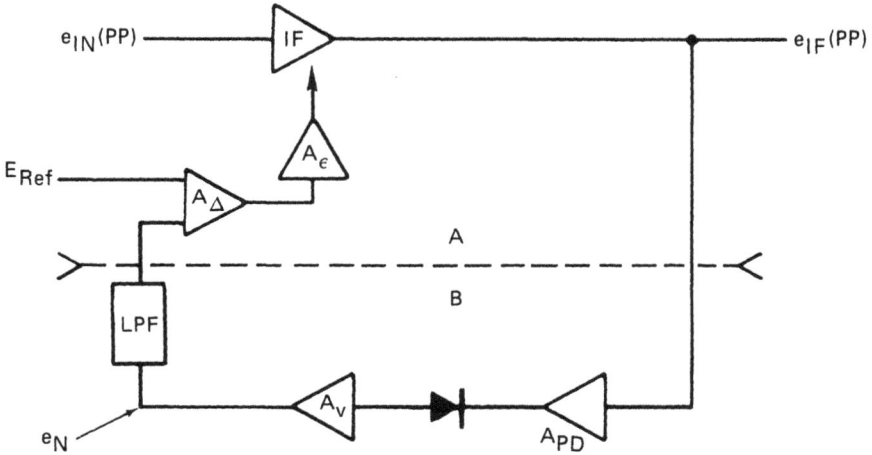

FIGURE 1-5.   Block Diagram of Feedback Amplifier.

The loop gain (for constant input power) is found by breaking the AGC input to the variable gain IF amplifier and modulating the AGC voltage. The loop gain is thus the modulated output voltage, $\Delta AGC'$, divided by the modulated AGC voltage, ΔAGC, as illustrated in Figure 1-6:*

$$LG = \frac{\Delta AGC'}{\Delta AGC} \qquad (1\text{-}17)$$

or

$$LG = A_{AGC}A_{PD}A_{D}A_{v}A_{\Delta}A_{\varepsilon} \qquad (1\text{-}18)$$

---

*The effect of the low-pass filter will be neglected.   Ideally this sets the frequency response of the loop.

where

A_{AGC} = dynamic AGC gain, $\Delta e_{IF}(PP)/\Delta AGC$ (V/V)

A_{PD} = predetector gain, $\Delta e_{PD}(PD)/\Delta e_{IF}(PP)$ (V/V)

A_D = dynamic detector gain, $\Delta e_D/\Delta e_{PD}(PP)$ (V/V)

A_v = video gain, $\Delta e_N/\Delta e_D$ (V/V)

A_ε = error gain $\Delta AGC'/\Delta e_\varepsilon$ (V/V)

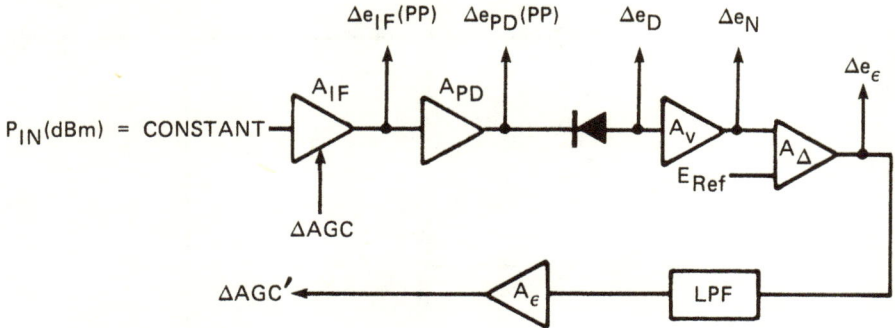

FIGURE 1-6. Method of Finding Loop Gain.

The dynamic AGC gain is a nonlinear quantity, but for small values of ΔAGC, may be given as (Appendix 1D)

$$A_{AGC} = 0.115 \, X e_{IF}(PP) \quad (V/V) \tag{1-19}$$

The dynamic detector gain is also nonlinear; however, it may be assumed linear over small values of $\Delta e_{PD}(PP)$. The dynamic detector gain for a square law detector (Appendix 1B) may be given as

$$A_{D,SL} = 5 \times 10^{-3} K_{SL} e_{PD}(PP) \quad (V/V) \tag{1-20}$$

where K_{SL} is a square law detector constant.

Substituting Equations (1-19) and (1-20) into (1-18),

$$LG_{SL} = \left[0.115 \, X \, e_{IF}(PP) \right] A_{PD} \left[5 \times 10^{-3} \right] K_{SL} e_{PD}(PP) A_v A_\Delta A_\varepsilon \tag{1-21}$$

10

However, since

$$e_{IF}(PP) = e_{PD}(PP)/A_{PD} \tag{1-22}$$

Equation (1-21) may be written as

$$LG_{SL} = 0.115\, X \left[e_{PD}(PP) \right]^2 5 \times 10^{-3} K_{SL} A_v A_\Delta A_\varepsilon \tag{1-23}$$

The detector output, e_D, is (Appendix 1B)

$$e_D = 2.5 \times 10^{-3} K_{SL} \left| e_{PD}(PP) \right|^2 \quad (V) \tag{1-24}$$

thus, under normalized conditions $e_D = e_{D,N}$

$$LG_{SL} = 0.23\, X A_v A_\Delta A_\varepsilon e_{D,N} \tag{1-25}$$

or , since

$$e_N = e_{D,N} \cdot A_v$$

$$LG_{SL} = 0.23 X A_\Delta A_\varepsilon e_N \tag{1-26}$$

Equation (1-26) is simple but very accurate. Using the same methods, but for a linear detector (Appendix 1B),

$$A_{D,Lin} = 1.59 \times 10^{-3} K_{Lin} \quad (V/V) \tag{1-27}$$

and

$$e_{D,Lin} = 1.59 \times 10^{-3} K_{Lin} e_{PD}(PP) \quad (V) \tag{1-28}$$

thus the loop gain becomes

$$LG_{Lin} = 0.115\, X A_\Delta A_\varepsilon e_N \tag{1-29}$$

It can be seen from Equations (1-5) and (1-6) that the loop gain and static regulation ($\Delta P_{IF}(dB)$) are dependent on $X A_\Delta A_\varepsilon e_N$; thus, for a given IF amplifier, $A_\Delta A_\varepsilon$ can be maximized to give the necessary loop gain,

11

$$\Delta P_{IF,SL}(dB) = 10 \log \left| \frac{\Delta P_{in}(dB)}{X A_\Delta A_\varepsilon e_N} + 1 \right| \tag{1-30}$$

and

$$LG_{SL} = 0.23 \, X A_\Delta A_\varepsilon e_N \tag{1-31}$$

Solving Equation (1-31) for $A_\Delta A_\varepsilon e_N$,

$$A_\Delta A_\varepsilon e_N = \frac{LG_{SL}}{0.23X} \tag{1-32}$$

and substituting into Equation (1-30),

$$\Delta P_{IF,SL}(dB) = 10 \log \left| \frac{0.23 \Delta P_{in}(dB)}{LG} + 1 \right| \tag{1-33}$$

For the linear detector this equation becomes

$$\Delta P_{IF,Lin}(dB) = 10 \log \left| \frac{0.12 \Delta P_{in}(dB)}{LG} + 1 \right| \tag{1-34}$$

Thus the loop gain uniquely determines the static regulation. Conversely, the static regulation uniquely determines the loop gain, as shown below.

Solving Equation (1-30) for $A_\Delta A_\varepsilon e_N$,

$$A_\Delta A_\varepsilon e_{N,SL} = \frac{\Delta P_{in}(dB)}{X \left(10^{\frac{\Delta P_{IF}(dB)}{10}} - 1 \right)} \tag{1-35}$$

Substituting Equation (1-35) into (1-31),

$$LG_{SL} = 0.23 \left[\frac{\Delta P_{in}(dB)}{\left(10^{\frac{\Delta P_{IF}(dB)}{10}} - 1 \right)} \right] \tag{1-36}$$

12

and for the linear detector,

$$LG_{Lin} = 0.115 \left| \frac{\Delta P_{in}(dB)}{\left(10^{\frac{\Delta P_{IF}(dB)}{20}} - 1\right)} \right| \qquad (1\text{-}37)$$

The dynamic regulation for the true integrator AGC loop of Figure 1-2b is similar to the low-pass-filter loop, except that the loop gain must be multiplied by the frequency-dependent gain of the integrator, A_{Int} (assuming a large low-frequency gain for the operational amplifier),

$$A_{Int} \simeq \frac{-Z_F}{R} \qquad (1\text{-}38)$$

where

$$Z_F = \frac{-j}{2\pi fC} \qquad (1\text{-}39)$$

or

$$A_{Int} \simeq \frac{0.159}{fRC} \angle + 90^\circ \qquad (1\text{-}40)$$

Thus,

$$LG_{SL} = \frac{0.036 \, XA_\Delta A_\varepsilon e_N}{fCR} \qquad (1\text{-}41)$$

and

$$LG_{Lin} = \frac{0.018 \, XA_\Delta A_\varepsilon e_N}{fCR} \qquad (1\text{-}42)$$

The frequency response for the low-pass-filter loop will be determined by the low-pass filter (assuming that the open loop frequency, neglecting the filter, is much larger than the filter frequency response, which is usually the case).

13

Loop Rise Time and Bandwidth (f_{3dB})

The loop rise time, τ_r, is defined as the 10 to 90% AGC response time to step changes in input power. The rise time is dependent on the nonlinear characteristics of the detector; thus, the equations presented (see Appendix 1E for the derivation) are valid only for small input steps (less than ± 2 dBm).

The rise times and bandwidth for the integrator and low-pass-filter loops are the same, and may be given as

$$\tau_{r,SL} = \frac{9.56 \, RC}{XA_\Delta A_\varepsilon e_N} \qquad f_{3dB,SL} = \frac{0.35}{\tau_{r,SL}} \qquad (1\text{-}43)$$

$$\tau_{r,Lin} = \frac{19.13 \, RC}{XA_\Delta A_\varepsilon e_N} \qquad f_{3dB,Lin} = \frac{0.35}{\tau_{r,Lin}} \qquad (1\text{-}44)$$

As can be seen, the loop rise time is also dependent on $XA_\Delta A_\varepsilon e_N$; however, the rise time can be calculated independently in terms of R and C.

The rise times given in Equations (1-43) and (1-44) are for CW input loops. What is the rise time for a pulse loop rather than a CW loop? The effect a pulse AGC loop has on τ_r is quite easy to determine. Figure 1-7 illustrates a basic pulse AGC loop, and the pertinent timing is shown in Figure 1-8.

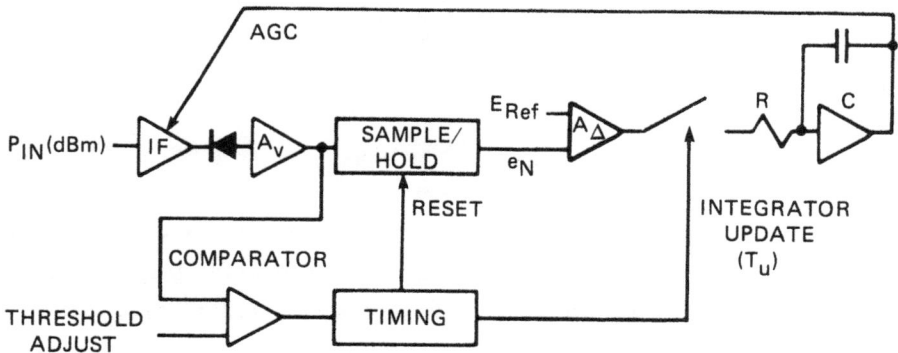

FIGURE 1-7. Basic Pulse AGC Block Diagram.

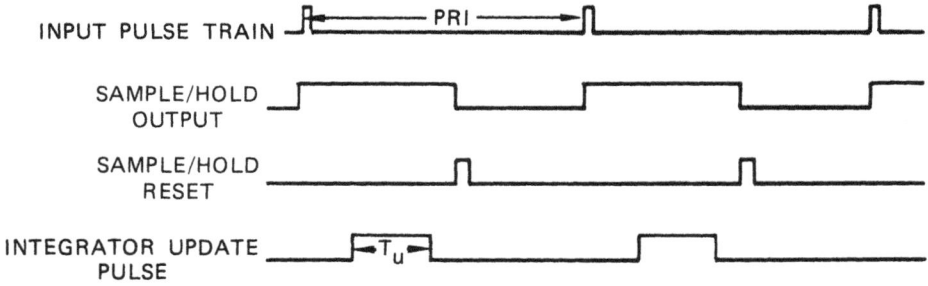

FIGURE 1-8. Basic Pulse AGC Timing.

The basic operation of the pulse AGC loop is straightforward; the input pulse is amplified, detected, sampled, and compared to the reference voltage, E_{Ref}. The amplified video also triggers a comparator (threshold) that initiates the timing. The integrator update switch is closed for a given period, T_u, every pulse; thus the integrator is only allowed to update the AGC loop during T_u. The effective time constant (T_{pulse}) for the updated integrator is

$$T_{pulse} = RC \left(\frac{PRI}{T_u} \right) \tag{1-45}$$

The update duty cycle, D, may be given as

$$D = \frac{T_u}{PRI} \tag{1-46}$$

Thus, Equation (1-45) may be written as

$$T_{pulse} = \frac{RC}{D} \tag{1-47}$$

15

All equations thus presented for loop rise time, ι_r, may now be given in more general terms as

$$\iota_{r,SL} = \frac{9.56\ RC}{X A_\Delta A_\varepsilon e_N D} \qquad f_{3dB,SL} = \frac{0.35}{\iota_{r,SL}} \qquad (1\text{-}48)$$

$$\iota_{r,Lin} = \frac{19.13\ RC}{X A_\Delta A_\varepsilon e_N D} \qquad f_{3dB,Lin} = \frac{0.35}{\iota_{r,Lin}} \qquad (1\text{-}49)$$

It should be noted that sampling the AGC loop makes it a sampled data system. This publication will assume that any input modulation frequency is much smaller (by at least ten) than the sampling rate (PRF in pulsed AGC). If this condition is not met, instability may well result.

Automatic Gain Control Design Verification

The preceding section presented the equations that characterize an AGC tracking loop. In this section, we will verify these equations using a simple AGC tracking loop. Nonideal parameters will be discussed (i.e., the IF amplifier's variable gain slope, X, is not linear over the full AGC range), and practical design equations will be presented with these nonideal parameters in mind.

Square Law Detector Test Circuit

Figure 1-9 illustrates the circuit used to validate the equations already presented. The philosophy here is to analyze an existing AGC loop to verify the equations rather than to verify by design (Chapter 2 presents the design for several practical AGC loops). The square law detector is discussed first, then the linear.

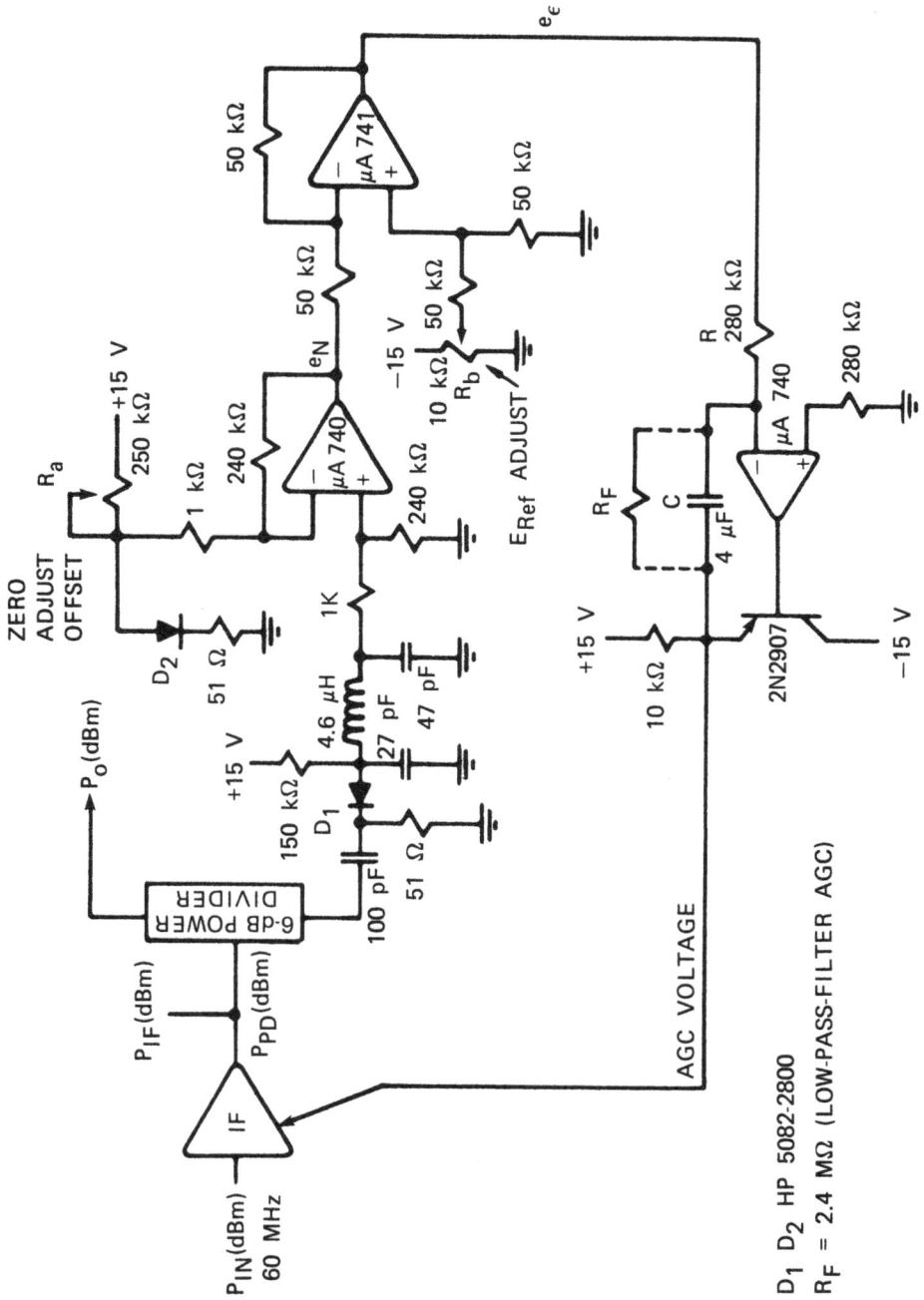

FIGURE 1-9. Square Law Detector AGC Test Circuit.

D_1 D_2 HP 5082-2800
R_F = 2.4 MΩ (LOW-PASS-FILTER AGC)

17

Figure 1-10 illustrates the square law characteristics for the detector used (see Appendix 1B for a discussion of detector characteristics). A normalized video reference, e_N, of -1 volt will be used; thus the detector output, e_D, is

$$e_D = e_N/A_v \tag{1-50}$$

or, since $A_v \simeq 240$,

$$e_D = 1/240 = 4.2 \, mV \tag{1-51}$$

The detector input power required to give an e_D of 4.2 mV is (Appendix 1B)

$$P_{PD}(dBm) = -10 \log \frac{K_{SL}}{e_D(mV)} \tag{1-52}$$

$$P_{PD}(dBm) = -10 \log \left(\frac{350}{4.2} \right) \tag{1-53}$$

or

$$P_{PD}(dBm) = -19.2 \, dBm \tag{1-54}$$

Detector D_2 is used as a temperature stabilizing element to minimize the dc offset effects of D_1. With no IF present, R_a is adjusted until $e_N = 0$ volt). Resistor R_b is adjusted to give the desired normalized video voltage, e_N (-1 volt). The -6 dB power splitter enables the output power variation to be monitored. The IF output power is

$$P_{IF}(dBm) = P_{PD}(dBm) + 6 \, dB \tag{1-55}$$

where both outputs of the power splitter are the same ($P_o(dBm) = P_{PD}(dBm)$). Resistor R_F is used for the low-pass-filter AGC.

1,000

100

DETECTOR OUTPUT (e_D), mV

10

1

0.1

0

-10

-20

-30

DETECTOR OUTPUT (e_D), dBV

-40

-50

-60

-70

-80

$$K_{SL} = (3.5)\ 10^{\frac{-(-20)}{10}} \left(\frac{mV}{dBm}\right)$$

$$K_{SL} = 350\ \frac{mV}{dBm}$$

3.5 mV

+15 V

-20 dBm

HP 5082-2800   150 kΩ

100 pF   4.7 μH

P_{PD}(dBm) e_D

51 Ω

27 pF   47 pF

f = 60 MHz

0   -5   -10   -15   -20   -25   -30   -35

DETECTOR INPUT P_{IN}, dBm

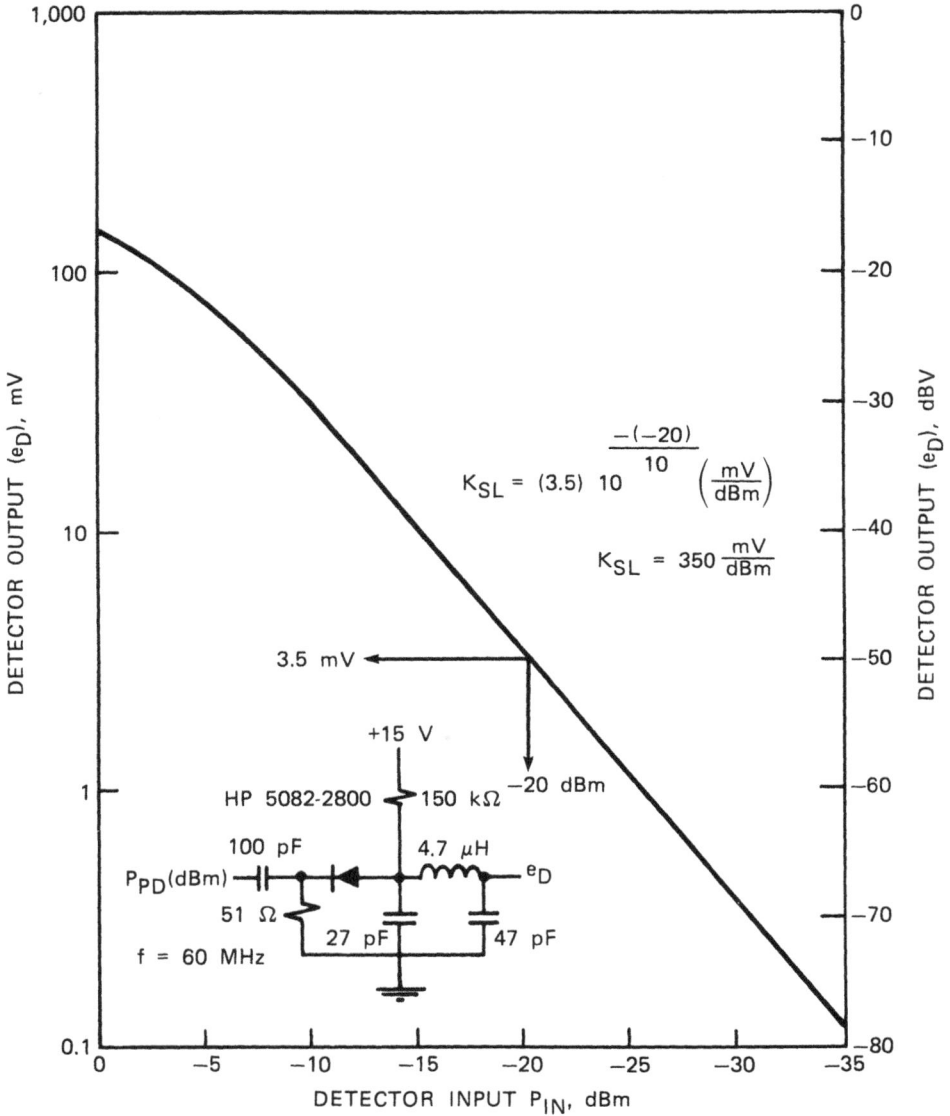

FIGURE 1-10. Detector Output Versus Input Power
(HP 5082-2800).

19

Figure 1-11 illustrates the variable gain characteristics of the IF amplifier. It will be noticed that the slope varies with AGC voltage (which is typical of many commercial variable gain IF amplifiers). The variation in X with AGC voltage and gain is given in Table 1-1. As can be seen, there is more than a three-to-one variation in X. Thus the loop gain, static regulation (for the low-pass-filter AGC), and rise time of the loop will be a function of AGC voltage and, thus, the input power, P_{in}(dBm).

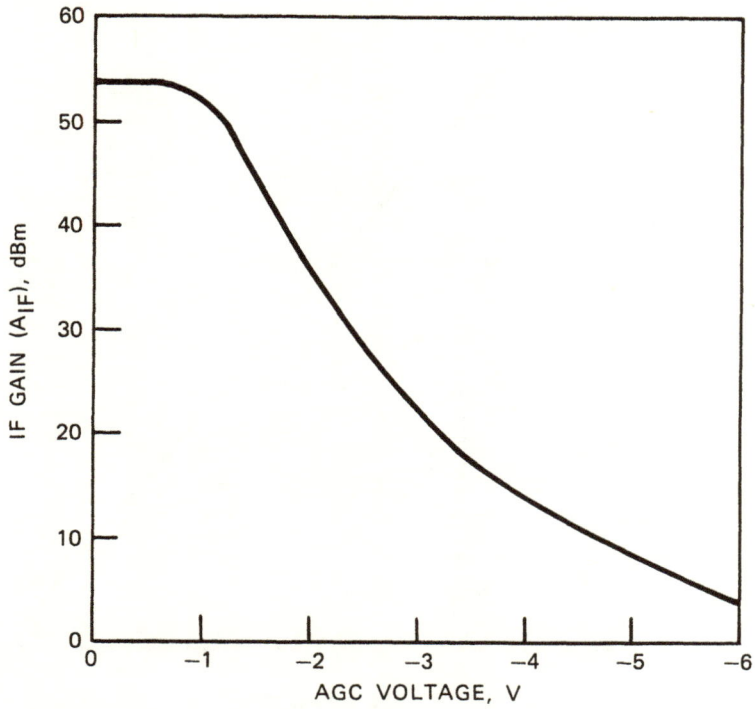

FIGURE 1-11. IF Amplifier Variable Gain Characteristics.

TABLE 1-1. IF Amplifier Variable Gain Slope.

AGC voltage, V	Gain, dB	X, dB/V
−1.75	40	18.2
−2.85	30	13.9
−3.25	20	9.1
−4.6	10	5.5

The static regulation for the low-pass-filter AGC was presented as (Equation (1-5))

$$\Delta P_{IF,SL}(dB) = 10 \log \left[\frac{\Delta P_{in}(dB)}{X A_\Delta A_\varepsilon e_N} + 1 \right] \tag{1-56}$$

which assumes a linear gain slope, X. Obviously the gain slope varies (Figure 1-11 and Table 1-1), and Equation (1-56) is not valid. This problem is easily corrected, however, by noting that

$$\Delta AGC = \frac{\Delta P_{in}(dB)}{X} \tag{1-57}$$

where ΔAGC is the total change in AGC voltage for the total desired input dynamic range, $\Delta P_{in}(dB)$. Equation (1-57) may now be written as

$$\Delta P_{IF,SL}(dB) = 10 \log \left[\frac{\Delta AGC}{A_\Delta A_\varepsilon e_N} + 1 \right] \tag{1-58}$$

which defines the static regulation for a practical low-pass-filter AGC loop using a square law detector (the linear detector loop will be presented shortly). The static regulation for the integrator loop will still be near zero because of the large static (dc) gain of the integrator.

The parameters for the low-pass-filter AGC loop may now be given as (see Figure 1-9)

$$A_{PD}(dB) = -6; P_{PD}(dBm) = -19.2; P_{IF}(dBm) = -13.2$$

$$e_N = -1 \text{ (adjusted at } -3 \text{ volts AGC)}; A_v = 240$$

$$A_\Delta = -1; A_\varepsilon = -8.57; R_F C = 9.6$$

The IF input was varied from -70 to -10 dBm ($\Delta P_{in}(dB) = 60$), and the AGC voltage varied from -1.01 to -6.87 volts ($\Delta AGC = 5.86$ volts), as shown in Figure 1-12. The output power varied from -22.2 dBm at an AGC voltage of -1.01 volts to -19.3 dBm at an AGC voltage of 6.87 volts ($\Delta P_o(dB) = 2.9$), as shown in Figure 1-13. Thus the change in output power, $\Delta P_o(dB)$, is

$$\Delta P_{o,SL}(dB) = 2.9 \tag{1-59}$$

The predicted change in output power ($\Delta P_o = \Delta P_{IF}$), using Equation (1-58), is

$$\Delta P_{o,SL}(dB) = 10 \log\left[\frac{5.86}{(-1)(-8.57)(1)} + 1\right] = 2.26 \, dB \tag{1-60}$$

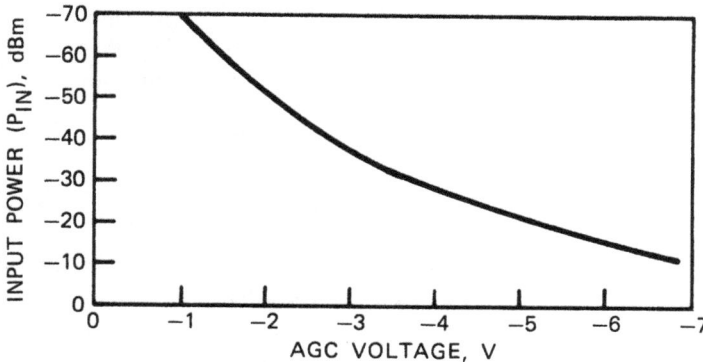

FIGURE 1-12. Input Power Versus AGC Voltage (Low-Pass Filter).

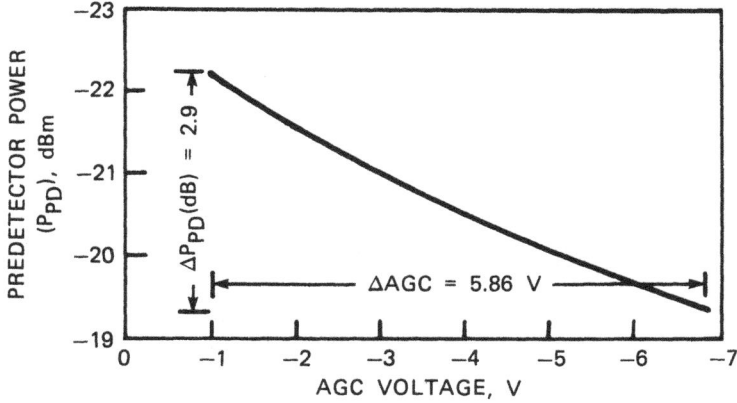

FIGURE 1-13. Detector Input Power Versus AGC Voltage (Low-Pass Filter).

and is in excellent agreement with the measured value of 2.9 dB. The normalized video voltage, e_N, varied from -0.78 volt at an AGC voltage of -1.01, to -1.42 volts at an AGC voltage of -6.87 volts. Thus the measured change in e_N, Δe_N(dB), is

$$\Delta e_{N,SL}(dB) = 20 \log (1.42/0.78) = 5.2 \tag{1-61}$$

The predicted value is twice the change in output power, since the change in e_D in decibels is twice the change in input power in decibels for a square law detector (see Appendix 1B). Thus,

$$\Delta e_{N,SL}(dB) = 20 \log \left(\frac{\Delta AGC}{A_\Delta A_\varepsilon e_N} + 1 \right) \tag{1-62}$$

or

$$\Delta e_{N,SL}(dB) = 20 \log \left| \frac{5.86}{(-1)(-8.57)(1)} + 1 \right| = 4.5 \tag{1-63}$$

which also is in excellent agreement with the measured value of 5.2 dB.

23

The value for e_N was changed to -2 volts by adjusting R_b, and the measured difference in ΔP_o(dB) for input variations from -70 to -10 dBm was

$$\Delta P_{o,SL}(dB) = 1.23 \tag{1-64}$$

The predicted value, using Equation (1-58), is

$$\Delta P_{o,SL}(dB) = 10 \log \left| \frac{5.86}{(1)(8.57)(2)} + 1 \right| = 1.28 \tag{1-65}$$

which again is in excellent agreement with the measured result. The measured change in e_N, $\Delta e_{N,SL}(dB)$, was 2.15 dB. The predicted change is in excellent agreement with the measured value, as shown in Equation (1-66):

$$\Delta e_{N,SL}(dB) = 20 \log \left| \frac{5.86}{(1)(8.57)(2)} + 1 \right| = 2.55 \tag{1-66}$$

The loop rise time was measured for input variations of ± 1, ± 5, and ± 10 dB at three input levels. Table 1-2 summarizes the results.

TABLE 1-2. Loop Rise Time Results.

P_{in}, dBm	AGC, V	X, dB/V	Loop rise time, τ_r, s					
			-1 dB	+1 dB	-5 dB	+5 dB	-10 dB	+10 dB
-53	-1.93	15	0.7	0.8	0.85	0.6	1.0	0.5
-37	-3.16	9.1	1.0	0.95	1.1	0.8	1.25	0.7
-26	-4.51	5.5	1.25	1.25	1.5	1.1	1.7	0.9

The equations predicting the loop rise time are valid only for small input deviations, as discussed in the preceding section and Appendix 1E. The rise time for the low-pass-filter loop was given as (Equation (1-43))

$$t_{r,SL} = \frac{9.56\ RC}{X A_\Delta A_c e_N} \tag{1-67}$$

The predicted values are compared to the measured values in Table 1-3 (± 1 dB), and as can be seen, there is good agreement.

TABLE 1-3. Measured and Predicted Loop Rise Times.

P_{in}, dBm	X, dB/V	τ_r (measured), s	τ_r (predicted - Eq. 1-67), s
-53	15.0	0.8	0.71
-37	9.1	1.0	1.18
-26	5.5	1.25	1.94

The deviation in loop rise time at large input variations (Table 1-2) is due to the nonlinear behavior of the detector. Two methods to minimize this effect are given later in the chapter.

The loop gain was measured as illustrated in Figure 1-6. The modulating frequency was well below the low-pass-filter bandwidth.

$$f_{3dBV}(LPF) \simeq \frac{1}{2\pi RC} \tag{1-68}$$

(The open loop frequency response for this circuit (with C removed) is in excess of 3 kilohertz; thus the $R_F C$ filter will determine the bandwidth for frequencies up to about 1 kilohertz.)

The measured loop gains are listed in Table 1-4. The predicted loop gain is (Equation (1-26))

$$LG_{SL} = 0.23 \, X A_\Delta A_\epsilon e_N \tag{1-69}$$

Table 1-4 also compares the measured and predicted loop gains, which can be seen to be quite close.

TABLE 1-4. Low Pass Filter Loop Gains.

P_{in}, dBm	AGC	X, dB/V	Measured loop gain	Predicted loop gain
−50	−1.95	15.0	25.7	29.6
−35	−3.19	9.1	15.0	17.9
−25	−4.53	5.5	12.8	10.4

Figure 1-9 (with R_F removed) illustrates the basic integrator AGC loop using a square law detector. Figure 1-14 illustrates the input power versus AGC voltage for $e_N = -1$ volt. The measured change in the output power, ΔP_o(dB), was 0.12 dB for inputs from −65 to −10 dBm. The predicted change, from Equation (1-12), is 0 dBm, which is in excellent agreement with the measured result.

Table 1-5 summarizes the loop rise time results.

The predicted rise time, again valid only for small input variations (Equation (1-67)), is given in Table 1-6, and agrees favorably with the measured results. The value for e_N was doubled, and the loop rise time was halved, as predicted by Equation (1-67).

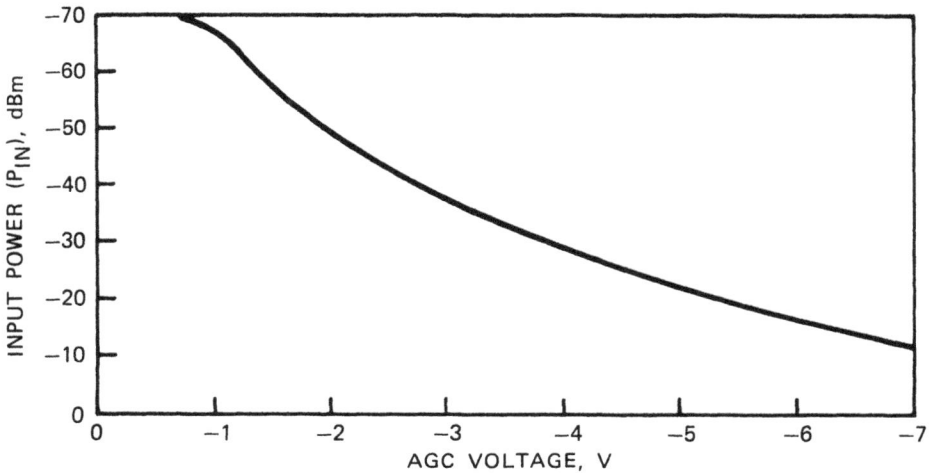

FIGURE 1-14. Input Power Versus AGC Voltage (Integrator, Square Law Detector).

TABLE 1-5. Integrator AGC Loop Rise Time Results.

P_{in},dBm	AGC, V	X, dB/V	Loop rise time, τ_r, s					
			-1 dB	+ 1 dB	-5 dB	+ 5 dB	-10 dB	+ 10 dB
− 50	1.95	15.0	0.58	0.55	0.68	0.4	0.7	0.4
− 35	3.19	9.1	0.95	0.95	1.1	0.9	1.2	0.6
− 25	4.53	5.5	1.5	1.4	1.8	1.2	1.9	0.9

TABLE 1-6. Measured and Predicted Integrator Loop Rise Times.

P_{in},dBm	X, dB/V	τ_r (measured), s	τ_r (predicted), s
− 50	15.0	0.58	0.71
− 35	9.1	0.95	1.17
− 25	5.5	1.5	1.94

27

The loop gain is frequency-sensitive, as discussed earlier.

$$LG_{SL} = \frac{0.036 \, X A_{\Delta} A_{\varepsilon} e_N}{fRC}$$ (1-70)

The measured loop gain for an input of -35 dBm ($X = 9.1$ dB/V) was 0.03 with an input frequency of 10 Hertz. The predicted value is

$$LG_{SL} = \frac{(0.036)(9.1)(1)(1)(1)}{(10)(280 \times 10^3)(4 \times 10^{-6})} = 0.0293$$ (1-71)

which is in excellent agreement with measured results.

Linear Detector Test Circuit

Figure 1-15 illustrates the circuit used to verify the linear detector AGC equations. The same variable gain amplifier used for the square law detector test circuit is used, as is the detector. Figure 1-16 illustrates the linear characteristics for the HP 5082-2800 detector. As can be seen, this detector has linear qualities above 0 dBm.

A normalized video reference, e_N, of -1 volt is used; the detector output, e_D, is

$$e_D = e_N/A_v$$ (1-72)

$A_v = 3.3$; thus,

$$e_D = -1/3.3 = -0.3 \text{ V}$$ (1-73)

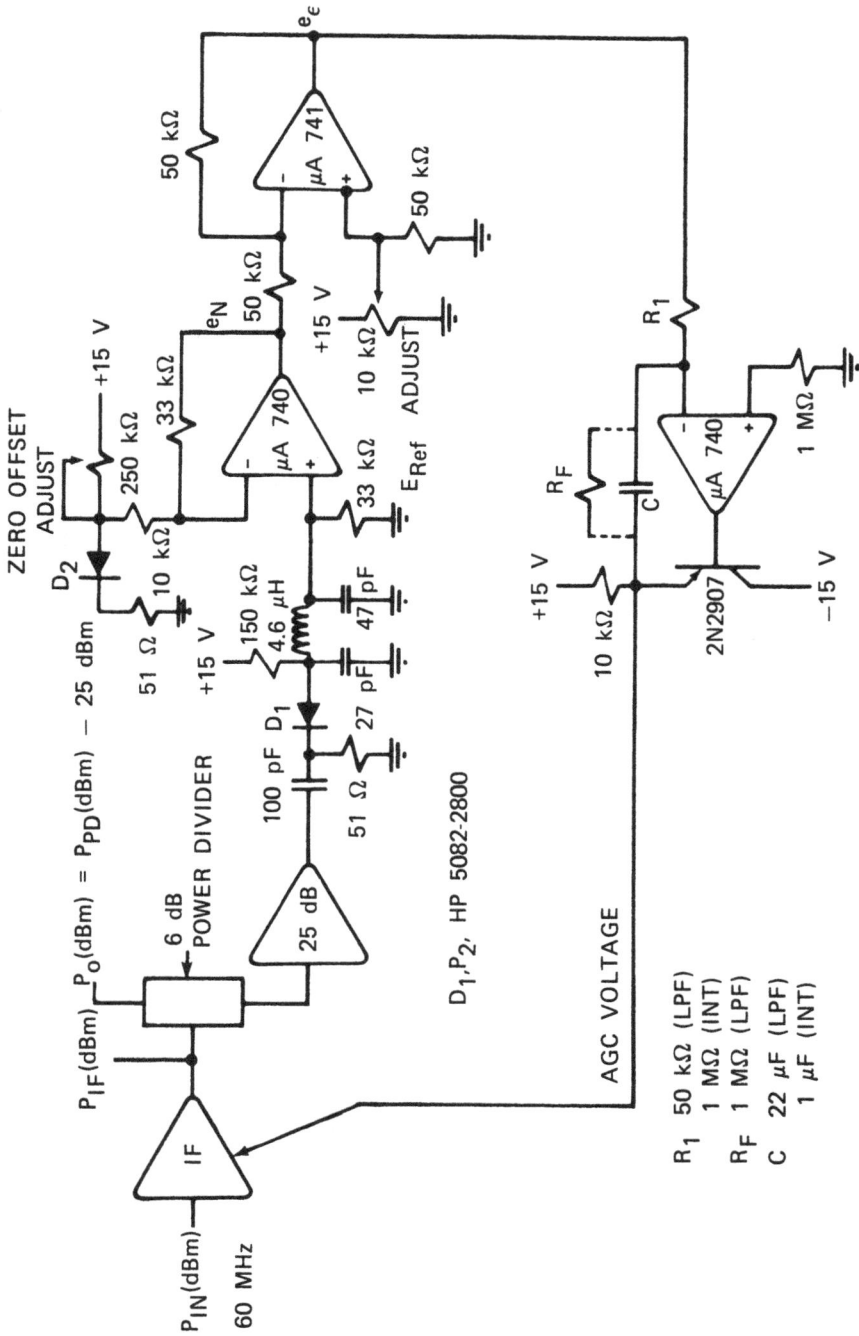

FIGURE 1-15. Linear Detector AGC Test Circuit.

29

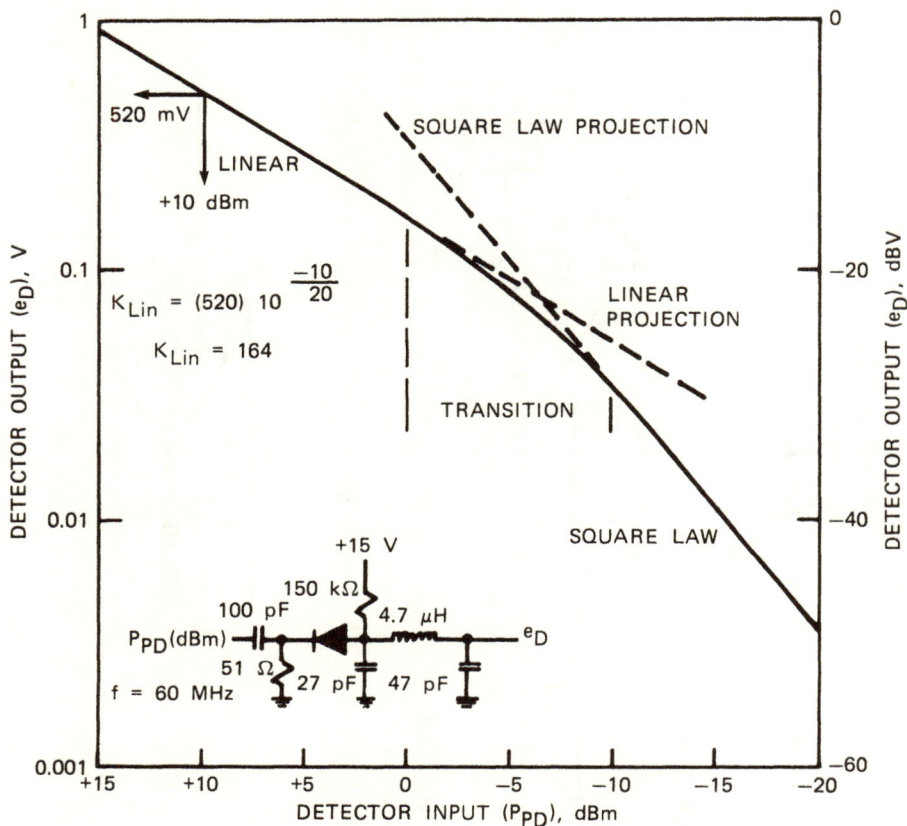

FIGURE 1-16. Detector Output Versus Input Power (HP 5082-2800).

The necessary input power to satisfy Equation (1-73) is (Appendix 1B)

$$P_{PD}(dBm) = -20 \log\left(\frac{K_{Lin}}{e_D(mV)}\right) \qquad (1-74)$$

30

$$P_{PD}(dBm) = -20 \log \left(\frac{164}{300} \right) \qquad (1\text{-}75)$$

or

$$P_{PD}(dBm) = 5.3 \, dBm \qquad (1\text{-}76)$$

and, referring to Figure 1-15,

$$P_{IF}(dBm) = 5.3 - 25 + 6 = -13.75 \, dBm \qquad (1\text{-}77)$$

The low-pass-filter AGC will be discussed first ($A_v = 3.3$, $A_\Delta = -1$, $A_\varepsilon = -20$, $e_N = -1$ volt). The static regulation may be given as

$$\Delta P_{o,Lin}(dB) = 20 \log \left(\frac{\Delta AGC}{A_\Delta A_\varepsilon e_N} + 1 \right) \qquad (1\text{-}78)$$

The input power was varied from -69 to 0 dBm ($\Delta P_{in}(dB) = 69$), and the AGC voltage varied from -0.82 to -7.03 volts ($\Delta AGC = 6.21$ volts), as shown in Figure 1-17. The output power from the 6-dBm power divider varied from -19.6 to -17.49 dBm ($\Delta P_o(dB) = 2.11$) over the AGC range, as shown in Figure 1-18. The nonlinearities for inputs larger than -10 dBm are due to the self-AGC effects of the input signal on the gain (the *signal* is large enough to have a controlling effect on gain). This condition must be avoided for linear operation (self-AGC will be discussed shortly). To avoid any effects of self-AGC, only inputs from -69 to -10 dBm will be used. Using Figures 1-17 and 1-18,

$$\Delta P_{in}(dB) = -59 \, dB \qquad (1\text{-}79)$$

$$\Delta AGC = 5.64 \, V \qquad (1\text{-}80)$$

$$\Delta P_o(dB) = 2.11 \, dB \qquad (1\text{-}81)$$

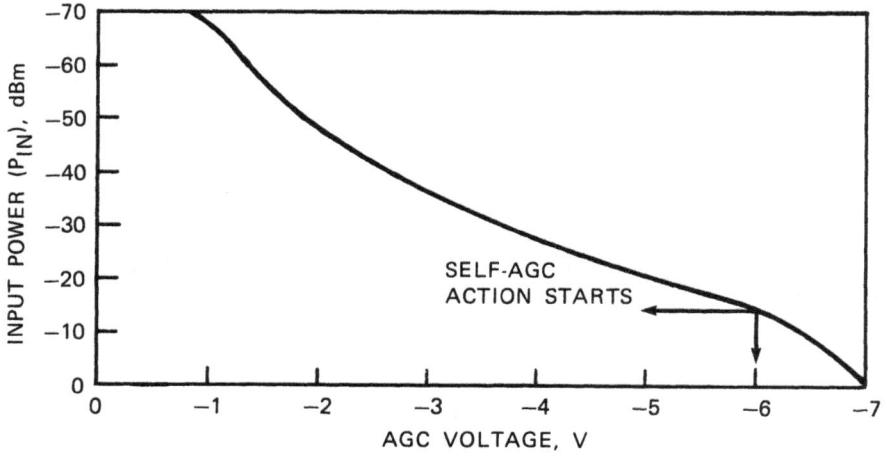

FIGURE 1-17. Input Power Versus AGC Voltage (LPF, Linear Detector).

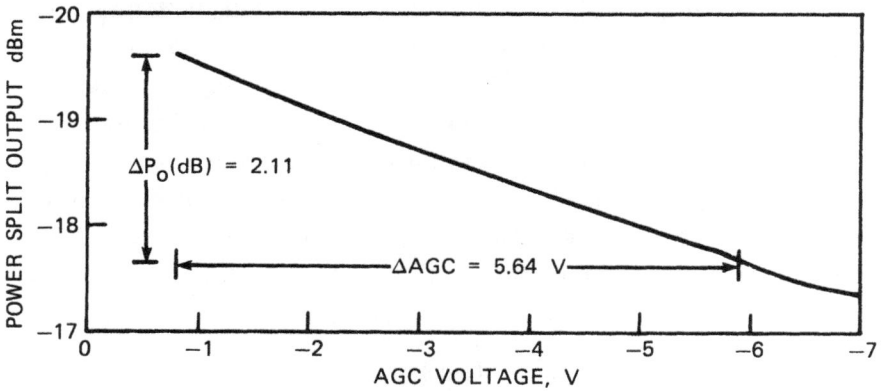

FIGURE 1-18. Power Splitter Output Power Versus AGC Voltage (LPF, Linear Detector).

The predicted change in output power is (Equation (1-78))

$$\Delta P_{o,Lin}(dB) = 20 \log \left[\frac{5.64}{(1)(-1)(-20)} + 1 \right] = 2.16 \, dB \qquad (1-82)$$

which is in excellent agreement with the measured value. The normalized video output, e_N, varied from -0.895 volt (AGC $= -0.82$ volt) to -1.143 volts (AGC $= -6.46$ volts). Thus the change in e_N is (in dB)

$$\Delta e_N(dB) = 20 \log (1.143/0.895) = 2.12 \, dB \qquad (1-83)$$

This is the same as the change in output power, which is to be expected for a linear detector. Thus

$$\Delta e_{N,Lin}(dB) = 20 \log \left(\frac{\Delta AGC}{A_\Delta A_\varepsilon e_N} + 1 \right) = \Delta P_o(dB) \qquad (1-84)$$

The loop rise time was measured for input variations of ± 1 dB ($AGC = 3.1$; $X = 9.1$ dB/V) and was 1.7 seconds. The predicted loop rise time is (Equation (1-44))

$$\iota_{r,Lin} = \frac{19.12 \, R_F C}{X A_\Delta A_\varepsilon e_N} \qquad (1-85)$$

or

$$\iota_{r,Lin} = \frac{19.12(1 \times 10^6)(22 \times 10^{-6})}{(9.1)(-1)(-20)(1)} = 2.31 \, seconds \qquad (1-86)$$

which is in good agreement with the measured value.

The loop gain was measured to be 22 ($AGC = -3.1$, $X = 9.1$ dB/V). The predicted loop gain is (Equation (1-29))

$$LG_{Lin} = 0.115 \, X A_\Delta A_\varepsilon e_N \qquad (1-87)$$

33

or

$$LG_{Lin} = 0.115(9.1)(-1)(-20)(1) = 20.93 \qquad (1\text{-}88)$$

which is in excellent agreement with the measured value.

The linear detector integrator AGC loop is illustrated in Figure 1-15, with R_F removed, $C = 1\ \mu F$, and $R_1 = 1\ M\Omega$. Figure 1-19 illustrates the input power versus AGC voltage. This curve also deviates from the expected at inputs larger than -10 dBm due to the self-AGC effects.

The power output from the 6-dB power divider varied from -18.3 dBm ($P_{PD}(\text{dBm}) = 6.7$ dBm) at -65 dBm input power to -18.53 dBm ($P_{PD}(\text{dBm}) = 6.47$ dBm) at -10 dBm input power. Thus the output power changed:

$$\Delta P_o(\text{dB}) = 6.7 - 6.47 = 0.23\ \text{dB} \qquad (1\text{-}89)$$

which is very close to the 0-dBm change predicted. The normalized video voltage, e_N, varied from -1.042 to -1.018 volts over the same input power range. Thus,

$$\Delta e_N(\text{dB}) = 20 \log \frac{1.042}{1.018} = 0.2\ \text{dB} \qquad (1\text{-}90)$$

which is similar to the change in output power, as is to be expected with the linear detector ($\Delta e_N(\text{dB}) = \Delta P_{PD}(\text{dB})$).

The loop rise time measured for an AGC voltage of -3.1 ($X = 9.1$ dB/V) at ± 1 dB input power deviation was 1.8 seconds. The predicted value (Equation (1-85)) is

$$\tau_r = \frac{19.12(1 \times 10^6)(1 \times 10^{-6})}{(9.1)(1)(-1)(-1)} = 2.1\ \text{seconds} \qquad (1\text{-}91)$$

and is in excellent agreement with the measured value.

34

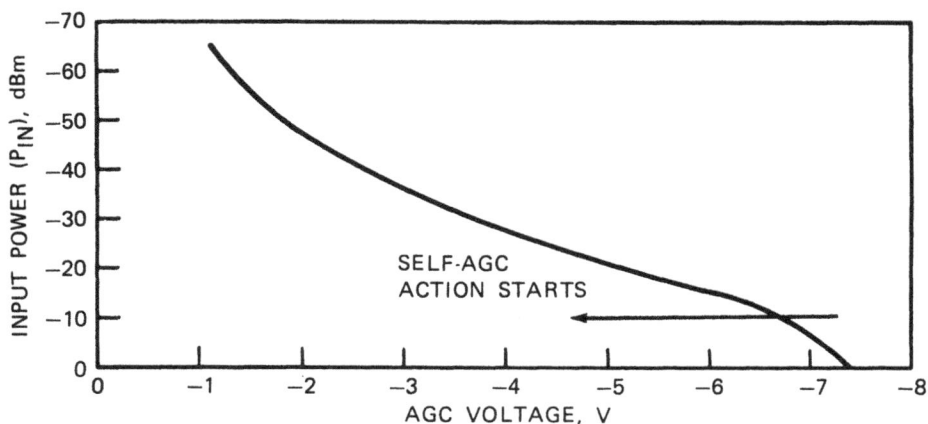

FIGURE 1-19. Input Power Versus AGC Voltage
(Integrator, Linear Detector).

The measured loop gain is 0.0147 for a modulation frequency of 10 Hertz. The predicted value (Equation (1-42)) is

$$LG_{Lin} = \frac{0.018 \, X A_{\Delta} A_{\varepsilon} e_N}{fRC} \qquad (1\text{-}92)$$

or

$$LG_{Lin} = \frac{0.018(9.1)(1)(1)(1)}{10(1 \times 10^6)(1 \times 10^{-6})} = 0.0164 \qquad (1\text{-}93)$$

which is also in excellent agreement with the measured value.

A series switch was placed between the differencing amplifier, A_{Δ}, and the integrator to verify Equations (1-48) and (1-49). A pulse repetition frequency of 1 kilohertz and update time, T_u, of 500 microseconds were used. The duty cycle is thus

$$D = \frac{T_u}{PRI} = T_u \, (PRF) \qquad (1\text{-}94)$$

35

or

$$D = (500 \times 10^{-6})(1 \times 10^{3}) = 0.5 \qquad (1\text{-}95)$$

The expected loop rise times are, from Equations (1-48) and (1-49),

$$t_{r,SL} = \frac{9.56\,RC}{X A_{\Delta} A_{\varepsilon} e_{N} D} \qquad (1\text{-}96)$$

$$t_{r,Lin} = \frac{19.12\,RC}{X A_{\Delta} A_{\varepsilon} e_{N} D} \qquad (1\text{-}97)$$

or the loop rise times should double for a duty cycle of 0.5. Measured results show exact agreement.

It should be noted that a holding capacitor (sample-hold circuit) is needed to hold the error voltage, e_{ε}, for the low-pass-filter AGC in pulse-AGC applications (Figure 1-7).

Figures 1-20, 1-21, and 1-22 summarize the AGC loop design equations for easy reference.

A basic understanding of various methods of controlling amplifier gain will now be presented.

Variable Gain Elements

This section will present a basic discussion of several modern variable gain techniques. The approach to vary IF/RF amplifier gain in years past (early 1960s) was to vary the operating point of the amplifying transistor (forward

36

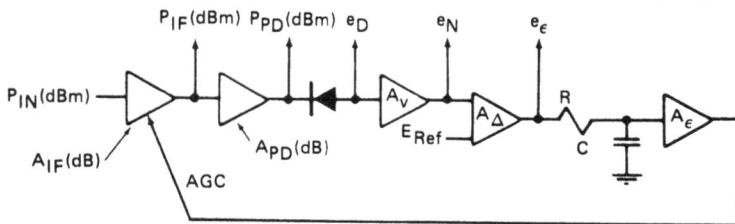

$$e_N = A_v e_D \ , \ \Delta e_N(dB) = 2\Delta P_{PD}(dB) \ , \ \Delta AGC = \frac{\Delta P_{in}(dB)}{X} \ , \ D = \text{low-pass filter update duty cycle}$$

STATIC REGULATION

$$\Delta P_{PD}(dB) = 10 \log \left[\frac{\Delta AGC}{A_v A_\Delta A_\varepsilon e_D} + 1 \right] \qquad A_v A_\Delta A_\varepsilon = \frac{\Delta AGC}{\left[10^{\frac{\Delta P_{PD}(dB)}{10}} - 1 \right] e_D}$$

$$\Delta P_{PD}(dB) = 10 \log \left[\frac{(\Delta AGC \times 10^3) 10^{\frac{-P_{PD}(dBm)}{10}}}{K_{SL} A_v A_\Delta A_\varepsilon} + 1 \right] \qquad A_v A_\Delta A_\varepsilon = \frac{(\Delta AGC \times 10^3) 10^{\frac{-P_{PD}(dBm)}{10}}}{K_{SL} \left[10^{\frac{\Delta P_{PD}(dB)}{10}} - 1 \right]} - 1$$

DYNAMIC REGULATION

$$LG = 0.23 \ X \ A_v A_\Delta A_\varepsilon e_D = (0.23 \times 10^{-3}) K_{SL} \ X \ A_v A_\Delta A_\varepsilon 10^{\frac{P_{PD}(dBm)}{10}}$$

$$A_v A_\Delta A_\varepsilon = \frac{4.35 \ LG}{X e_D} = \frac{(4.35 \times 10^3) LG \ 10^{\frac{-P_{PD}(dBm)}{10}}}{K_{SL} X} \qquad LG = \frac{0.23 \Delta P_{in}(dB)}{10^{\frac{\Delta P_{PD}(dB)}{10}} - 1}$$

LOOP RISE TIME AND BANDWIDTH* (f_{3dB})

$$t_r = \frac{2.2 \ RC}{(LG) D} = \frac{9.56 \times 10^3 \ RC \left[10^{\frac{-P_{PD}(dBm)}{10}} \right]}{K_{SL} X A_v A_\Delta A_\varepsilon D} \qquad f_{3dB} \approx \frac{0.35}{t_r} = \frac{(0.16)(LG)D}{RC}$$

* It is assumed that the loop frequency response is determined by the low-pass filter.

FIGURE 1-20. Low-Pass Filter, Square Law Detector Design Summary.

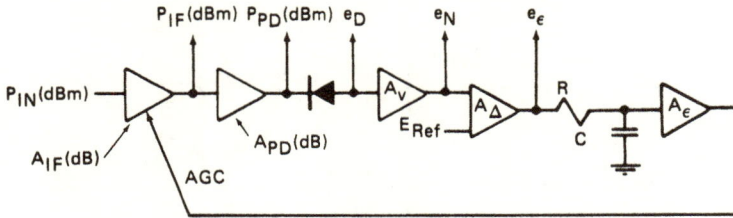

$$e_N = A_v e_D \ , \ \Delta e_N (dB) = \Delta P_{PD}(dB), \ \Delta AGC = \frac{\Delta P_{in}(dB)}{X} \ , \ D = \text{low-pass filter update duty cycle}$$

STATIC REGULATION

$$\Delta P_{PD}(dB) = 20 \log \left[\frac{\Delta AGC}{A_v A_\Delta A_\varepsilon e_D} + 1 \right] \qquad A_v A_\Delta A_\varepsilon = \frac{\Delta AGC}{\left[10^{\frac{\Delta P_{PD}(dB)}{20}} - 1 \right] e_D}$$

$$\Delta P_{PD}(dB) = 20 \log \left[\frac{(\Delta AGC \times 10^3)10^{\frac{-P_{PD}(dBm)}{20}}}{K_{Lin} A_v A_\Delta A_\varepsilon} + 1 \right] \qquad A_v A_\Delta A_\varepsilon = \frac{(\Delta AGC \times 10^3)10^{\frac{-P_{PD}(dBm)}{20}}}{K_{Lin} \left[10^{\frac{\Delta P_{PD}(dB)}{20}} - 1 \right]}$$

DYNAMIC REGULATION

$$LG = 0.115 \, X A_v A_\Delta A_\varepsilon e_D = (0.115 \times 10^{-3}) X K_{Lin} A_v A_\Delta A_\varepsilon 10^{\frac{P_{PD}(dBm)}{20}}$$

$$A_v A_\Delta A_\varepsilon = \frac{8.7 \, LG}{X e_D} = \frac{(8.7 \times 10^3) LG \, 10^{\frac{-P_{PD}(dBm)}{20}}}{K_{Lin} X} \qquad LG = \frac{0.115 \, \Delta P_{in}(dB)}{10^{\frac{\Delta P_{PD}(dB)}{20}} - 1}$$

LOOP RISE TIME AND BANDWIDTH* (f_{3dB})

$$t_r = \frac{2.2 \, RC}{(LG) D} = \frac{19.1 \times 10^3 \, RC \left[10^{\frac{-P_{PD}(dBm)}{20}} \right]}{X K_{Lin} A_v A_\Delta A_\varepsilon D} \qquad f_{3dB} \cong \frac{0.35}{t_r} = \frac{(0.159)(LG)D}{RC}$$

* It is assumed that the loop frequency response is determined by the low-pass filter.

FIGURE 1-21. Low-Pass Filter, Linear Detector Design Summary.

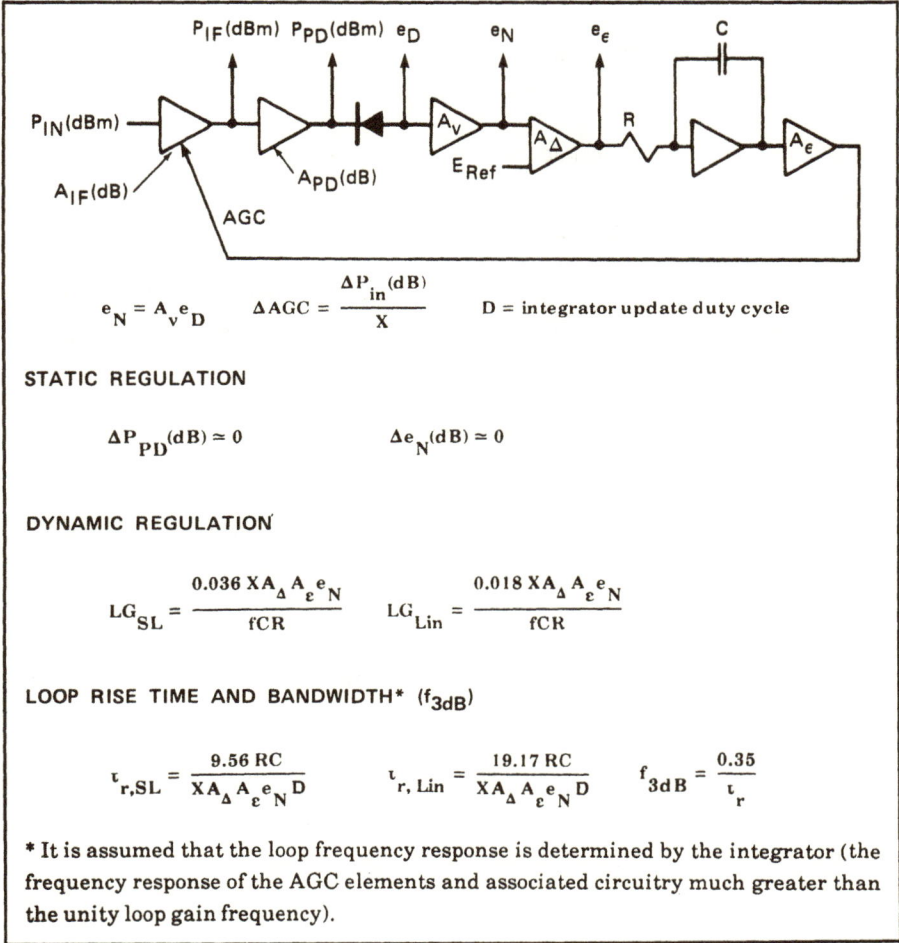

$$e_N = A_v e_D \qquad \Delta AGC = \frac{\Delta P_{in}(dB)}{X} \qquad D = \text{integrator update duty cycle}$$

STATIC REGULATION

$$\Delta P_{PD}(dB) \simeq 0 \qquad\qquad \Delta e_N(dB) \simeq 0$$

DYNAMIC REGULATION

$$LG_{SL} = \frac{0.036 \, X A_\Delta A_\varepsilon e_N}{fCR} \qquad\qquad LG_{Lin} = \frac{0.018 \, X A_\Delta A_\varepsilon e_N}{fCR}$$

LOOP RISE TIME AND BANDWIDTH* (f_{3dB})

$$\iota_{r,SL} = \frac{9.56 \, RC}{X A_\Delta A_\varepsilon e_N D} \qquad \iota_{r,Lin} = \frac{19.17 \, RC}{X A_\Delta A_\varepsilon e_N D} \qquad f_{3dB} = \frac{0.35}{\iota_r}$$

* It is assumed that the loop frequency response is determined by the integrator (the frequency response of the AGC elements and associated circuitry much greater than the unity loop gain frequency).

FIGURE 1-22. Integrator Design Summary.

and reverse AGC). This method has two drawbacks that become serious in modern high-performance systems:

1. The dynamic range of the transistor is altered as the operating point changes.

2. Changes in the transistor's frequency characteristics occur, which make phase and amplitude tracking most difficult.

In general, there exists an operating point at which a transistor (Bipolar or FET) exhibits its best performance (low noise, bandwidth, gain, etc.), and any deviation from this operating point will result in degraded performance.

What, then, can the designer do if a variable gain is desired without changing the bias of the amplifying transistor? The answer, obviously, is to change something else.

P-N Junction Diodes as Variable Gain Elements

If a diode is included in either the emitter or collector of a transistor, changing the diode voltage or current will vary its impedance, and thus change the circuit's gain without affecting the transistor itself. Furthermore, depending upon whether the diode is placed in the emitter or collector leg, and whether it is controlled by a current or a voltage, four different gain versus control-signal transfer characteristics can be obtained: linear, logarithmic, hyperbolic, and inverse logarithmic.

The first step in implementing this scheme is to determine the dynamic resistance of the diode as a function of its applied voltage or current. The voltage across a forward-based diode is given by

$$V_F = \eta V_T \text{Ln}\, (I_F/I_s) + I_F R_s \tag{1-98}$$

where

V_F = diode forward voltage drop

η = diode constant (temperature insensitive)
  $\eta \simeq 2$ for silicon
  $\eta \simeq 1$ for Schottky

V_T = KT/q

K = Boltzmann's constant

T = absolute temperature

q = electron charge

I_F = diode forward current

I_s = reverse saturation current (temperature sensitive)

R_s = ohmic (bulk) resistance

If Equation (1-98) is differentiated with respect to I_F, the dynamic resistance (r_d) is obtained

$$r_d = \frac{\eta V_T}{I_F} + R_s \qquad (1\text{-}99)$$

or, as a function of forward voltage,

$$r_d = \frac{\eta V_T}{I_s} \exp\left(\frac{-V_F}{\eta V_T}\right) + R_s \qquad (1\text{-}100)$$

For modern diodes, the bulk resistance, R_s, is on the order of 1 ohm and can be neglected in both equations.

To make use of Equations (1-99) and (1-100), the two constants I_s and η must be determined. I_s is found by examining a plot of log I_F versus V_F (Figure 1-23) and extrapolating the straight portion (logarithmic region) until it intersects the ordinate. Alternatively, the reciprocal slope, m, of the straight portion can be measured and then, using any voltage current pair, V_x and I_x, I_s is given by

$$\log I_s = (m \log I_x - V_x)/m \qquad (1\text{-}101)$$

41

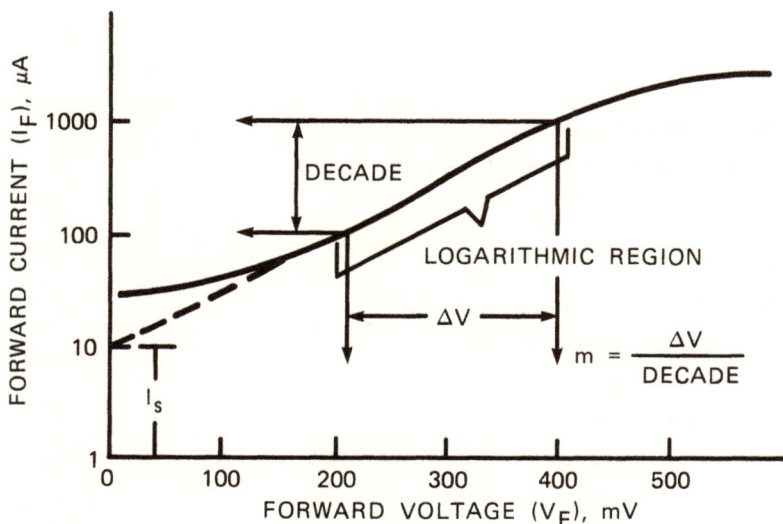

Y-axis: FORWARD CURRENT (I_F), μA — values 1, 10, 100, 1000

Labels on graph: DECADE, LOGARITHMIC REGION, ΔV, I_s

$$m = \frac{\Delta V}{DECADE}$$

X-axis: FORWARD VOLTAGE (V_F), mV — values 0, 100, 200, 300, 400, 500

HOW TO CALCULATE I_s: THE RECIPROCAL SLOPE, m, OF THE STRAIGHT PORTION OF THIS LOG I VS. V_F CURVE IS SEEN TO BE 0.2 VOLT PER DECADE. FOR A V_x OF, SAY, 0.4 VOLT, I_x IS 10^{-3} A, HENCE LOG I_x = −3. SUBSTITUTING THESE NUMBERS INTO EQUATION (1-101), A VALUE OF −5 IS FOUND FOR LOG I_s, HENCE, I_s = 10 μA AS VERIFIED BY THE EXTRAPOLATED (DASHED) LINE.

FIGURE 1-23. Calculation of I_s.

The slope, m, is measured in volts per decade of current. It should be noted that the curve will change with temperature.

I_s is a strong function of temperature while η is relatively unaffected by it. Furthermore, it is found that η is quite constant for a whole diode family, although I_s may vary greatly from unit to unit.

Most amplifier gains can be expressed in the form $A_v = R_C/R_E$ where R_C and R_E are collector and emitter resistances, respectively. Therefore, a variable-resistance diode placed across one of these gain-determining resistors may be used to control the amplifier's gain.

The four basic configurations shown in Figure 1-24 illustrate the flexibility of the technique [2].

One problem that this diode technique shares with more conventional methods of gain control is the undesired changing of the circuit gain by the gain-controlled signal itself (self-AGC). If the ac signal appearing across the gain-controlling diode is too large, it will strongly affect the circuit gain.

While this effect cannot be eliminated, it can be kept within set limits by making sure that the ac signal across the diodes does not exceed

$$e_d = m \log (A_v/A_x) \qquad (1\text{-}102)$$

where

e_d = the (instantaneous) value of the ac signal voltage across the diode

m = the reciprocal slope discussed in Figure 1-23

A_v = the desired gain of the amplifier

A_x = the minimum acceptable gain including the self-AGC effect.

Consider, as an example, a gain-controlled amplifier biased to provide a voltage gain of 10 (20 dB). Suppose that loss of 1 dB due to self-AGC is the maximum that can be tolerated; this would reduce the voltage gain to 8.9 (19 dB). If the diode slope, m, equals 120 mV/decade, then

$$e_d = 0.12 \log \frac{10}{8.9} = 6 \text{ mV} \qquad (1\text{-}103)$$

Thus the maximum ac signal across the diode should not exceed 6 mV.

43

LOGARITHMIC FUNCTION

A_V

50
10
1.0
0.1
0.01

V_{CC}
R_C
OUT
IN
R_E
DIODE SATURATION
CONTROL VOLTAGE
EFFECT OF R_E

$A_V = R_C / R_E'$
$R_C = 1k$
$R_E \gg r_d$

0.1 0.2 0.3 0.4 0.5 0.6
CONTROL VOLTAGE (V)

INVERSE LOG FUNCTION

A_V

200
100
10
1.0
0.1

EFFECT OF R_C
V_{CC}
OUT
CONTROL VOLTAGE
IN
R_E
DIODE SATURATION

$A_V = -R_C'/R_E$
$R_E = 100$
$R_C \gg r_d$

0.1 0.2 0.3 0.4 0.5 0.6
CONTROL VOLTAGE (V)

HYPERBOLIC FUNCTION

A_V

100
10
1.0
0.1

EFFECT OF R_C
V_{CC}
R_C
OUT
CONTROL CURRENT
R_E
DIODE SATURATION

$A_V = -R_C'/R_E$
$R_E = 100$ OHMS
$R_C \gg r_d$

10 10^2 10^3 10^4
CONTROL CURRENT (μA)

LINEAR FUNCTION

A_V

30
25
20
15
10
5
0

DIODE SATURATION
V_{CC}
R_C
OUT
IN
CONTROL CURRENT
R_E

$A_V = -R_C/R_E'$
$R_C = 1k$
$R_E \gg r_d$

0 0.5 1.0 1.5 2.0
CONTROL CURRENT (mA)

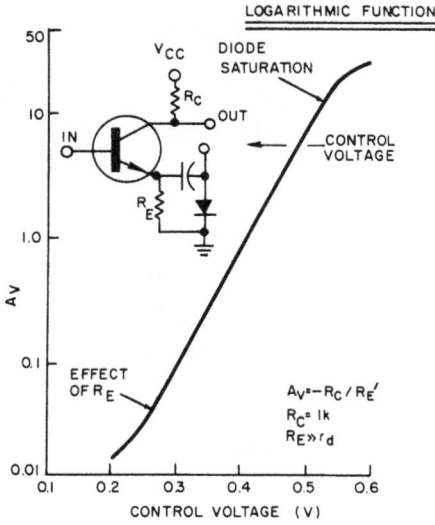

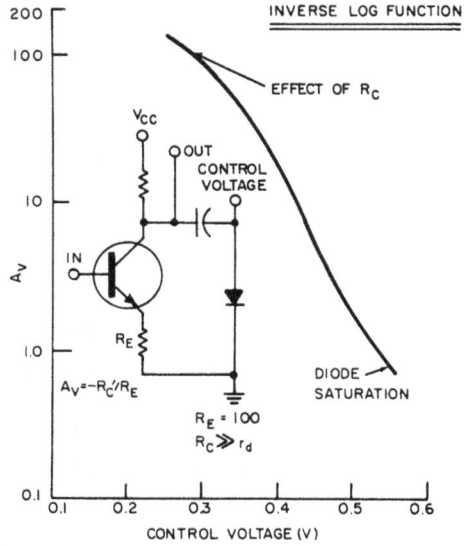

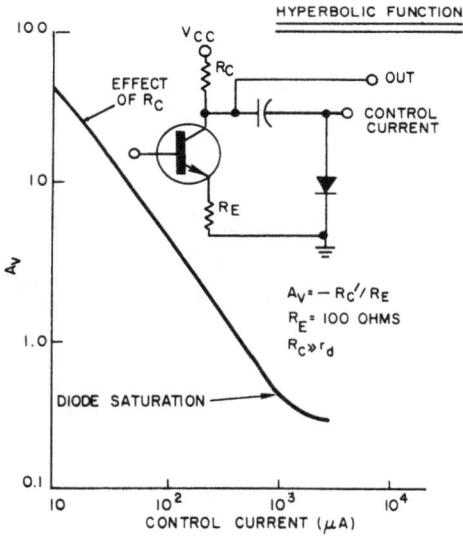

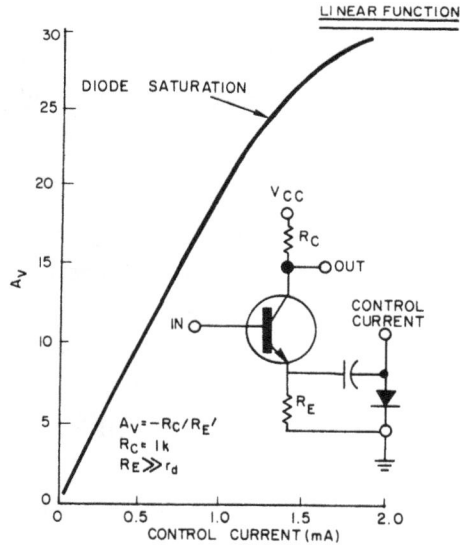

FIGURE 1-24. Four Basic P-N Diode AGC Configurations.

44

It is often necessary to have several amplifiers track in gain and phase. If the variable gain control is a current, little problem in diode matching for amplitude tracking arises.

The Four Basic P-N Diode AGC Configurations

The four circuits shown in Figure 1-24 demonstrate the flexibility of the diode gain-control approach.

The gain equations all assume a transistor alpha of unity. Deviations from the ideal response at low values of r_d are caused by diode saturation. At the other end of the scale, deviations arise as r_d approaches the value of the bypassed resistance.

For the cases of the emitter-coupled diode (linear and logarithmic functions) $R_E{}'$ is the parallel combination of R_E and r_d. When the diode is effectively across the collector resistor (hyperbolic and inverse log functions) $R_C{}'$ represents the parallel combinations of R_C and r_d.

A good working estimate of the gain temperature coefficient for the voltage-controlled cases is $+2$ mV/°C. This means that, for decreasing temperatures, increasing the applied voltage by 2 mV/°C will keep the gain constant. For the current-controlled cases, the temperature coefficient is approximately -0.17%/°C. Thus, for decreasing temperatures, decreasing I by 0.17%/°C will provide constant gain. (These numbers assume silicon diodes.)

Equation (1-99) indicates that the diode dynamic resistance, r_d, when driven from a current source, is not a function of I_s. Since η, for a given diode family, is quite constant and the diodes can be colocated to ensure they are at the same temperature, little problem in diode matching is encountered; simply ensure the diode slopes, m, meet the following criterion:

$$\frac{m_A}{m_B} = 10^{\frac{dB\,match}{20}} \tag{1-104}$$

Thus, to ensure a 1-dB gain match, the diode slopes (ΔV_F(mV)/*Decade I_F*) must be

$$\frac{m_A}{m_B} = 10^{\frac{1}{20}} = 1.12 \tag{1-105}$$

or the slopes must be matched to within 11.2%, which is quite easily accomplished.

If the control signal is a voltage, however, the diode resistance will depend on I_s, which is most temperature sensitive and varies from diode to diode. To ensure a given match, the diodes' forward voltage must be matched to within a specified limit, ΔV_F.

A simple equation relating ΔV_F for a given gain match may be given as

$$\Delta V_F \simeq (\eta T)(dB\,Match) \times 10^{-5} \tag{1-106}$$

Consider that two amplifiers must gain track to within 1 dB. Assuming Schottky diode ($\eta \simeq 1$) the necessary forward voltage match, ΔV_F, as a function of temperature, is given in Table 1-7.

To ensure proper phase tracking, the variable gain amplifiers must have a much wider bandwidth than the IF center frequency, and the variable diode

TABLE 1-7. ΔV_F Match Necessary
for a 1-dB Tracking Error.

Temperature, °C (°K)	ΔV_F, mV
+100 (373)	3.7
+27 (300)	3.0
−50 (223)	2.23

46

resistance should not have any affect on any frequency- or bandwidth-determining elements. A simple method to minimize any frequency or bandwidth degradation is to use variable attenuators following broadband fixed-gain stages as illustrated in Figure 1-25.

To ensure that, in an AGC loop, the low-pass filter or integrator (Figure 1-2) determines the frequency response, all capacitors must be small to ensure that the variable attenuation response is much faster than the desired AGC response time.

Figure 1-26 illustrates a block diagram of a commercial 60-MHz variable gain IF amplifier using P-N junction-diode controlled attenuators. Figure 1-27 illustrates the gain versus AGC voltage, and, as can be seen, the slope is fairly constant for gains from 10 to 55 dB. The self-AGC (for a 1-dB gain compression), as a function of gain and input power, is illustrated in Figure 1-28. The importance of this curve is that it clearly illustrates that a **maximum** output of −15 dB is permissible for a gain of 0 dB. Thus, over the full variable gain range of 60 dB, this amplifier does not have the output capability to drive a detector into its linear region. Thus the amplifier is suited only for square law detectors.

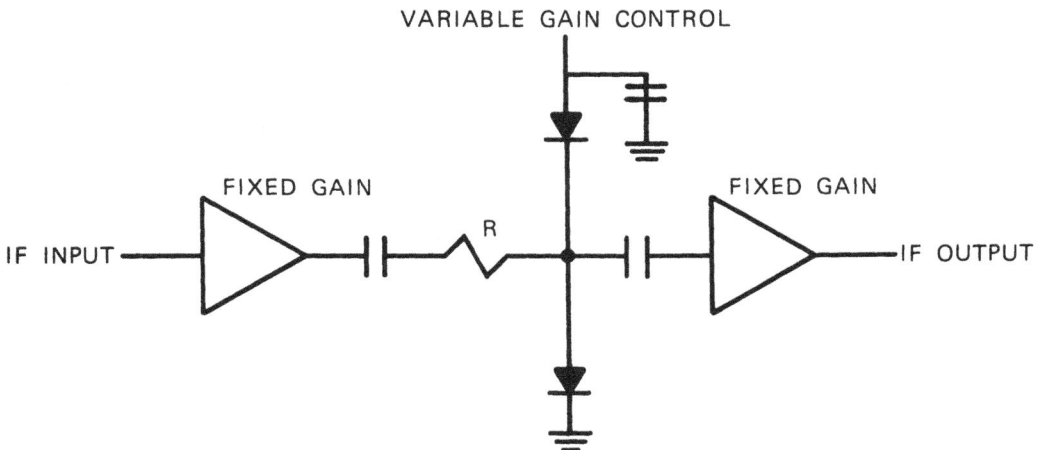

FIGURE 1-25. Basic Variable Attenuator Configuration.

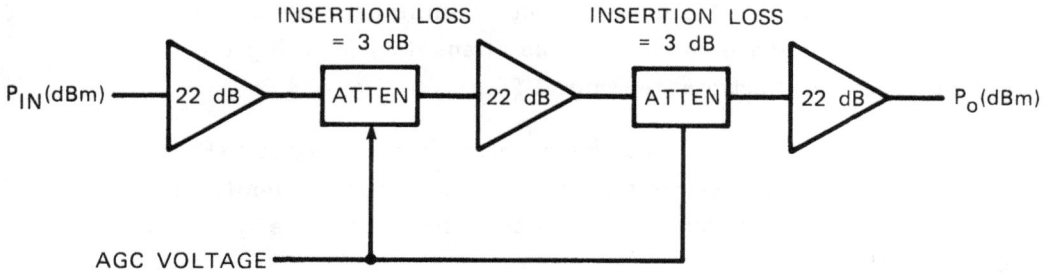

FIGURE 1-26. IF Amplifier Configuration.

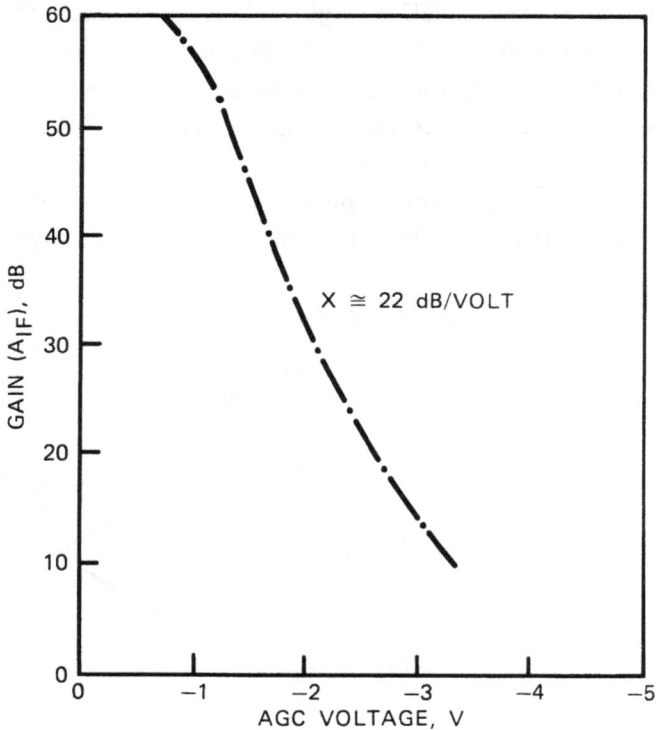

FIGURE 1-27. IF Amplifier Variable Gain Characteristics (Commercial 60-MHz Variable Gain Amplifier).

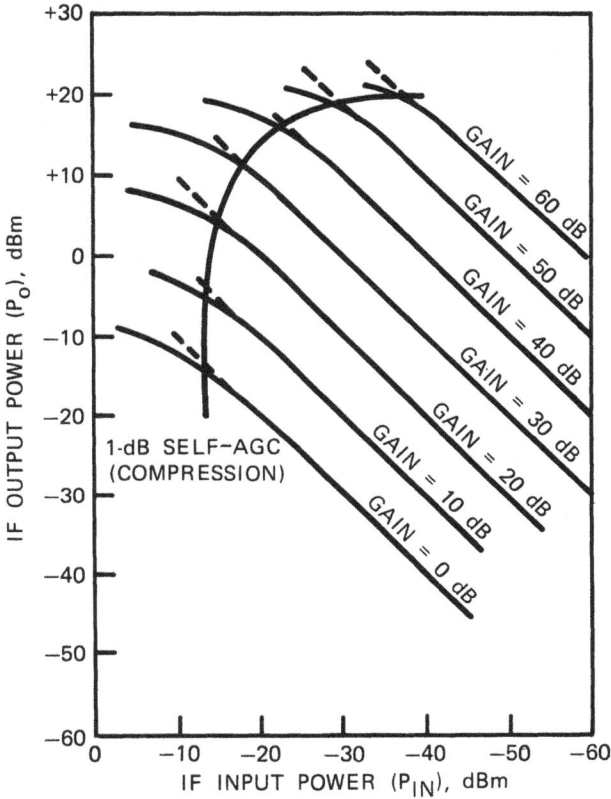

FIGURE 1-28. Self-AGC Characteristics
(Commercial 60-MHz Variable Gain
Amplifier).

P-N junction diodes have two serious limitations in modern EW systems:
(1) self-AGC starts at low input voltage levels, and (2) they are generally
frequency-limited to IFs of less than 100 MHz. The PIN diode, however,
overcomes these limitations.

PIN Diodes as Variable Gain Elements

The PIN diode appears as an almost pure resistance at frequencies up to
several thousand MHz, and is similar to the P-N junction diode in that this

resistance is a function of bias current or voltage. The general circuit concepts discussed earlier are valid for PIN diodes; however, they are most used as variable resistance elements in variable attenuators, usually in the π, T, or bridged T configuration, as illustrated in Figure 1-29. The designers of voltage variable attenuators using PIN diodes ensure that the 50 Ω input and output impedances are fairly insensitive to attenuation level.

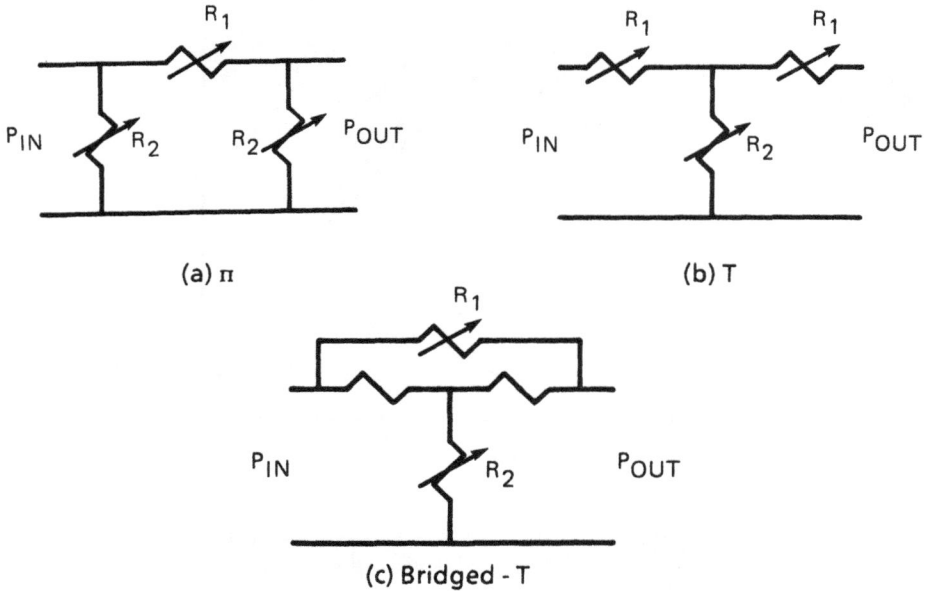

FIGURE 1-29. Several PIN Diode Configurations (R_1 and R_2 Represent the PIN's Variable Resistance).

The variable resistance of PIN diodes behaves in a manner similar to that of P-N junction diodes, in that it is an exponential function of voltage and a linear function of current. Also, the variable resistance is a strong function of temperature when driven from a voltage source, and relatively constant when driven from a current source.

50

Figure 1-30 illustrates the variable resistance versus forward current for a commercial PIN diode suited for use in variable attenuators [3, 4, and 5]. The variable resistance, r_{PIN}, may be given as

$$r_{PIN} = \frac{K_{P1}}{I_F^{K_{P2}}} \qquad (1\text{-}107)$$

where K_{P1} and K_{P2} are PIN diode constants and can be determined as illustrated in Figure 1-31.

The ease of using Figure 1-31 will become apparent by finding r_{PIN} for the "typical" resistance characteristic illustrated in Figure 1-30:

$$I_{FA} = 10\mu A \quad r_{PINA} \simeq 900\Omega \quad I_{FB} = 1\ mA, r_{PINA} \simeq 15\Omega$$

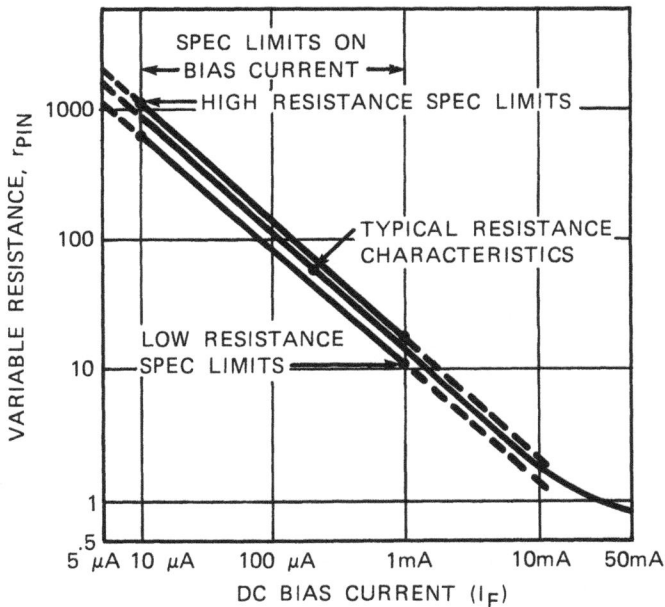

FIGURE 1-30. Variable Resistance
Characteristics for an HP 5082-3004
PIN Diode.

51

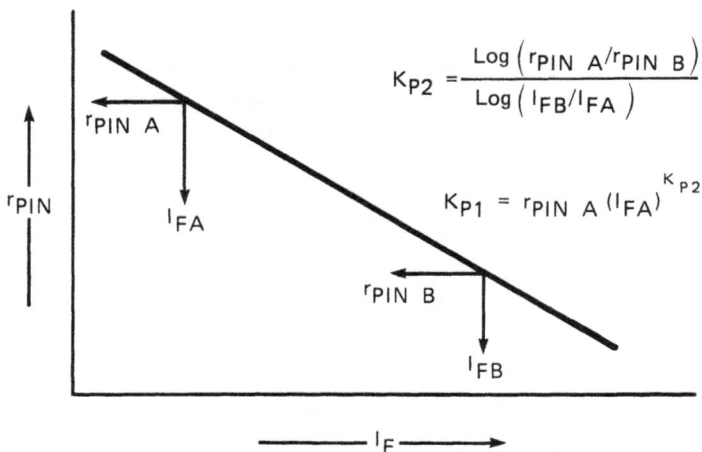

FIGURE 1-31. Method for Finding PIN Diode Constants.

$$K_{P2} = \frac{\text{Log } (900/15)}{\text{Log } \left(\dfrac{1 \times 10^{-3}}{10 \times 10^{-6}} \right)} = 0.889 \tag{1-108}$$

$$K_{P1} = (900)(10 \times 10^{-6})^{0.889} = 0.032 \tag{1-109}$$

Thus the variable resistance for the "typical diode" is

$$r_{PIN} = \frac{0.032}{I_F^{\,0.889}} \tag{1-110}$$

Figure 1-32 illustrates the basic configuration for a commercial π attenuator using PIN diodes. The variable attenuation characteristics are illustrated in Figures 1-33 and 1-34. Figure 1-35 illustrates the self-AGC effects, and as will be noted, there is little self-AGC for input powers less than 0 dBm. The PIN diode is far superior to P-N junction diodes with respect to self-AGC.

52

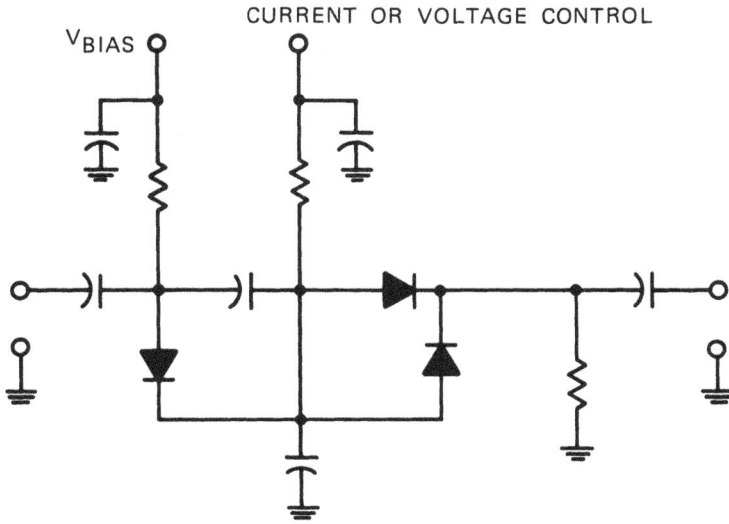

FIGURE 1-32. WJ G-1 PIN Attenuator.

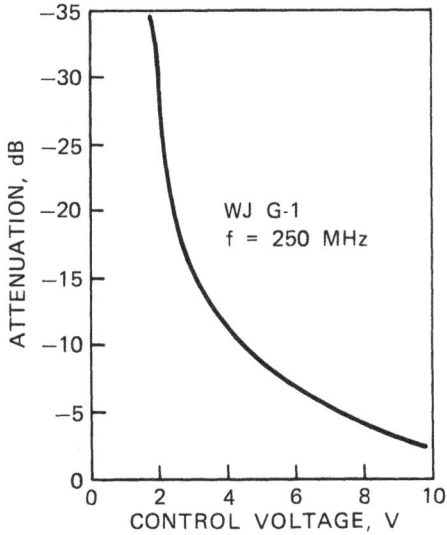

WJ G-1
f = 250 MHz

FIGURE 1-33. Voltage Drive
Characteristics for PIN Diode
Attenuator.

53

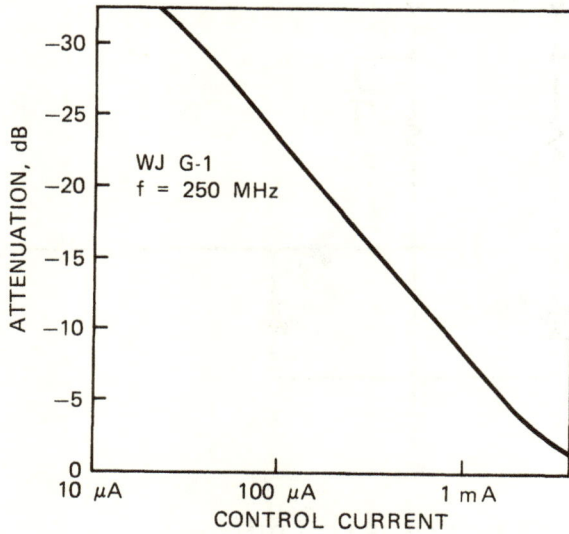

FIGURE 1-34. Current Drive Characteristics for PIN Diode Attenuator.

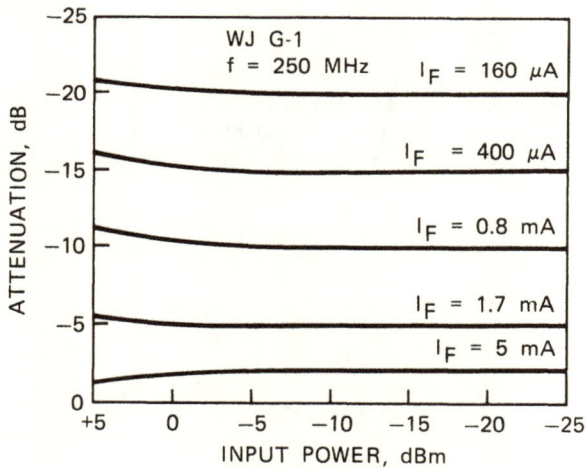

FIGURE 1-35. Attenuation Versus Input Power (Self-AGC Effect).

54

Figure 1-36 illustrates the effect of frequency on the variable attenuation characteristics of the PIN diode attenuator. The importance of this figure is that it is an indicator of reactive (rather than pure resistance) effects. For frequencies below 1,000 MHz, Figure 1-36 indicates that attenuations of less than 20 dB are due to the PIN diode resistance. This insensitivity to frequency greatly eases the amplitude and phase tracking from unit to unit. Amplitude tracking to within ±1 dB, and phase tracking to within less than ±5 degrees, is not difficult to obtain if the attenuation is limited to less than 20 dB and is via current control.

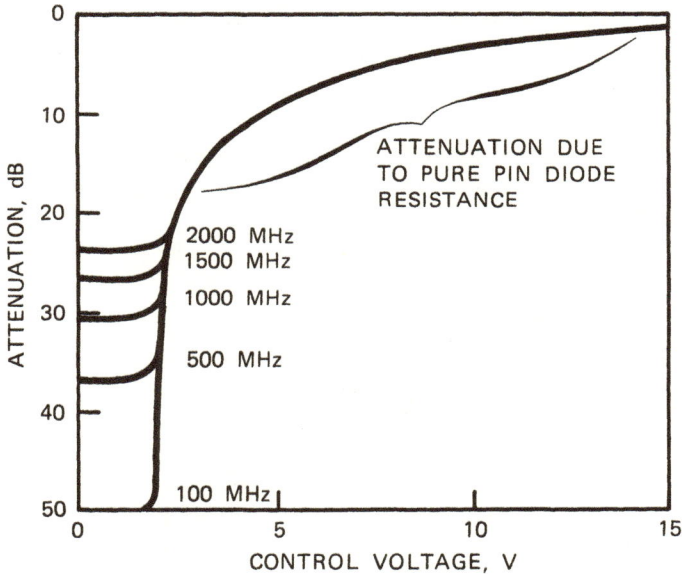

FIGURE 1-36. Attenuation Versus Control
Voltage as a Function of Frequency
(Obtained from the WJ CG-1 Data Sheet).

GaAs FETs as Variable Gain Elements

The advent of GaAs monolithic microwave integrated circuits (MMIC) extends the variable pure resistance needed for frequency insensitive variable attenuators to more than 10 GHz. Attenuator configurations using GaAs

FETs are similar to those illustrated in Figure 1-29; however, unlike PIN diodes, the FETs are monolithically integrated and very close matching would be expected.

Figure 1-37 [6] illustrates a modern voltage variable attenuator utilizing GaAs MMIC technology. The attenuator and operational amplifier within the box compare a reference attenuator that ensures a 50 Ω input and output impedance independent of attenuation. Assuming close matching between the reference and signal attenuators, the RF input sees a 50 Ω impedance (with RF output terminated in 50 Ω) independent of attenuation. The linearizing circuit compensates for the nonlinear attenuation versus control voltage (FETs are voltage-controlled, not current-controlled, devices). Figure 1-38 illustrates the variable attenuation characteristics.

Self-AGC effects depend upon FET design and geometry. Maximum input power of 0 dBm is typical (for 0.5-dB change in attenuation due to self-AGC).

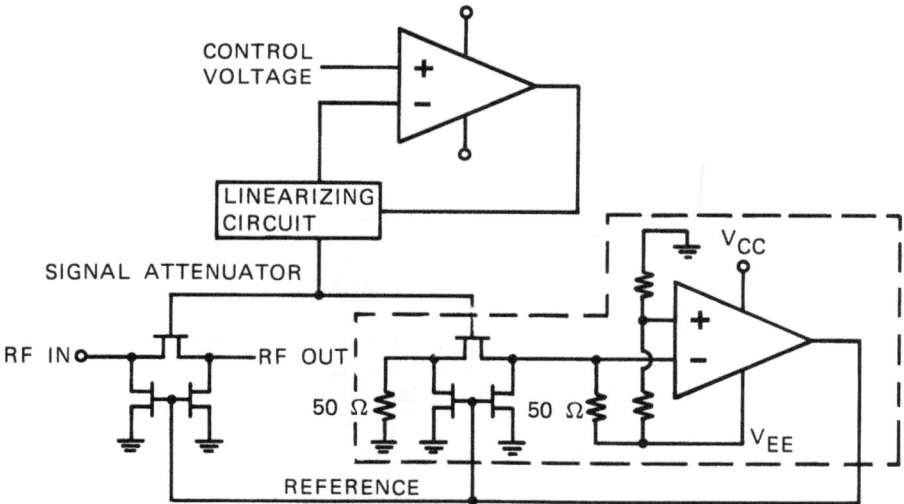

FIGURE 1-37. WJ-RG45 GaAs FET Voltage-Controlled Attenuator.

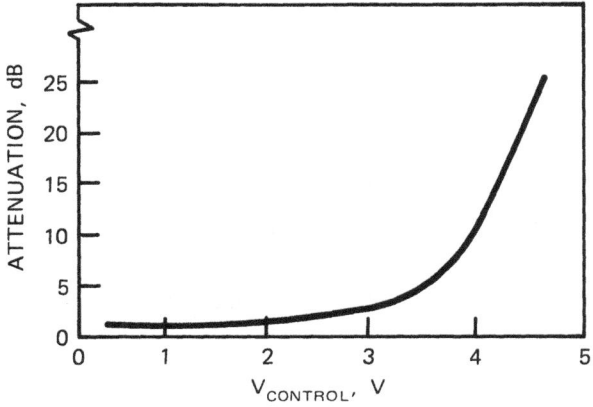

(a) Without Linearizing Circuit.

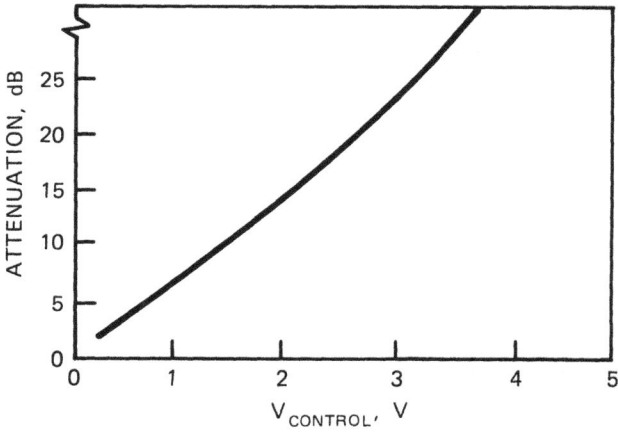

(b) With Linearizing Circuit.

FIGURE 1-38. WJ-RG45 Variable
Attenuation Characteristics.

Matching amplitude and phase is not as simple as one would expect for a monolithic design; in fact, it is easier to match PIN diode attenuators (amplitude less than ±1 dB, phase less than ±5 degrees) than GaAs FET attenuators. This may change as GaAs MMIC technology matures.

Many modern EW receivers are designed using commercially available wide-bandwidth fixed-gain amplifiers (which can be chosen for optimum noise figure, output power, etc.) and variable attenuators that can be positioned to optimize the linear dynamic range. The desired bandwidth and center frequency can be obtained with properly placed filters. A typical monopulse receiver employing variable gain may well appear as shown in Figure 1-39. Amplitude and phase matching can be accomplished by testing at the component level.

Applications

Several applications will now be presented to illustrate the design concepts previously described. Only the basic circuits will be presented since a

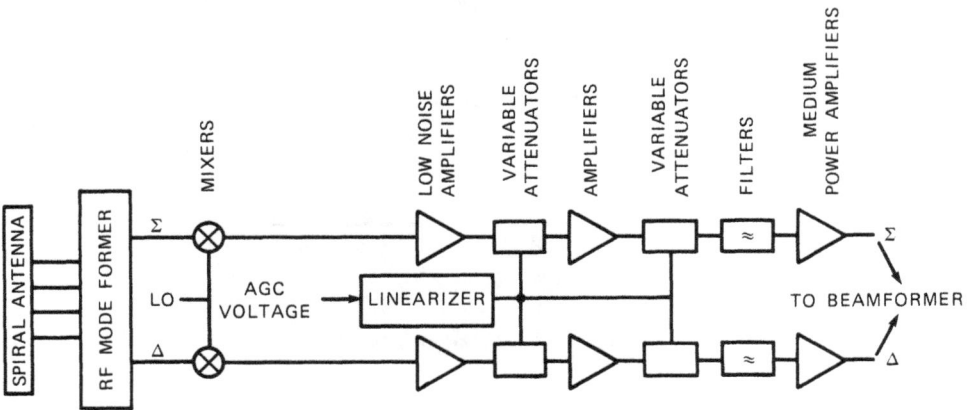

FIGURE 1-39. Basic Monopulse Receiver.

complete design, for even a simple AGC loop, would require more in-depth linear circuit design background than can be presented here.

To illustrate the design procedure, we will first consider a simple 30-MHz CW IF amplitude nulling loop.

A 30-MHz CW AGC loop is desired that will give a -15 dBm (± 0.25 dB) output. The loop is to operate in the square law region of the detector diode, and inputs to the IF amplifier are -60 dBm minimum and -10 dBm maximum. The loop should be of the low-pass filter design, with a closed loop bandwidth of at least 5 kHz (or loop rise time of 70 μs).

The variable gain characteristics for the 30-MHz amplifier is illustrated in Figure 1-40, and, as can be seen, the variable gain slope is nearly constant at $X = 70$ dB/V.

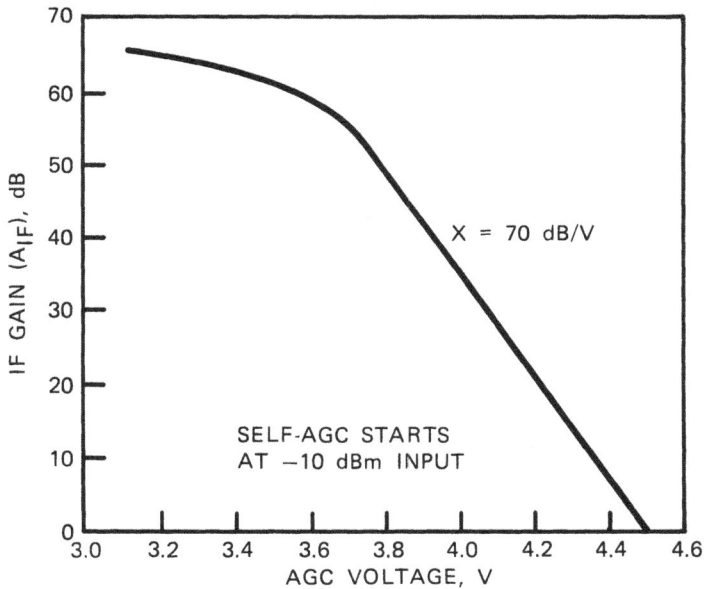

FIGURE 1-40. IF Gain Versus AGC Voltage (30-MHz IF Amplifier).

59

A back diode [7] will be used because of its near zero dc offset, thus minimizing the need for temperature stabilization. The diode constant for the detector (Ge BD-2) is

$$K_{SL} = 500 \text{ mV/mW} \tag{1-111}$$

Figure 1-41 illustrates the AGC loop and A_ε, A_Δ, and A_v have been combined to minimize parts. Since an output power of -15 dBm is desired, the power divider ensures the detector is also at -15 dBm (this power is within the detector's square law region). The IF amplifier's output will be 6-dB above the output power, $P_{IF} = -9$ dBm, which is well within the power output capabilities of the amplifier.

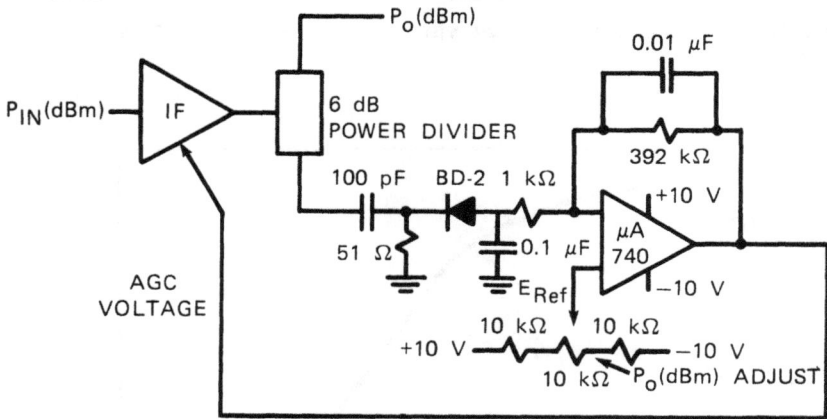

FIGURE 1-41. Single Operational Amplifier 30-MHz AGC Loop.

Referring to Figure 1-20, the value for $A_v A_\Delta A_\varepsilon$ may be found

$$A_v A_\Delta A_\varepsilon = \frac{(\Delta P_{in}(dB) \times 10^3) \, 10^{\frac{-P_o(dBm)}{10}}}{\Delta P_o(dB) \times K_{SL} \left[10^{\frac{\Delta P_o(dB)}{10}} - 1 \right]} \tag{1-112}$$

60

or

$$A_v A_\Delta A_\varepsilon = \frac{50 \times 10^3 \; 10^{\frac{-(-15)}{10}}}{(70)(500) \left| 10^{\frac{0.5}{10}} - 1 \right|}$$ (1-113)

Thus,

$$A_v A_\Delta A_\varepsilon = 370$$ (1-114)

We will use a gain of 392 (392 kΩ feedback resistor, 1 kΩ series resistor). The feedback capacitor needed for a loop rise time of 70 μs (5 kHz closed loop bandwidth) may be given as (Figure 1-20)

$$C = \frac{\iota_r \, X \, K_{SL} A_v A_\Delta A_\varepsilon D}{9.56 \times 10^3 \, R \left(10^{\frac{-P_o (dBm)}{10}} \right)}$$ (1-115)

or

$$C = \frac{(70 \times 10^{-6})(70)(500)(392)(1)}{9.56 \times 10^3 \, (392 \times 10^3) \, 10^{\frac{-(-15)}{10}}}$$ (1-116)

and

$$C = 0.0081 \; \mu F \, (\text{let } C = 0.01 \; \mu F)$$ (1-117)

E_{Ref} in Figure 1-41 is adjusted for $P_o = -15$ dBm at an input power of -35 dBm. Figure 1-42 illustrates the AGC voltage versus input power; inputs larger than -10 dBm have a controlling effect on the IF amplifiers' gain (self-AGC).

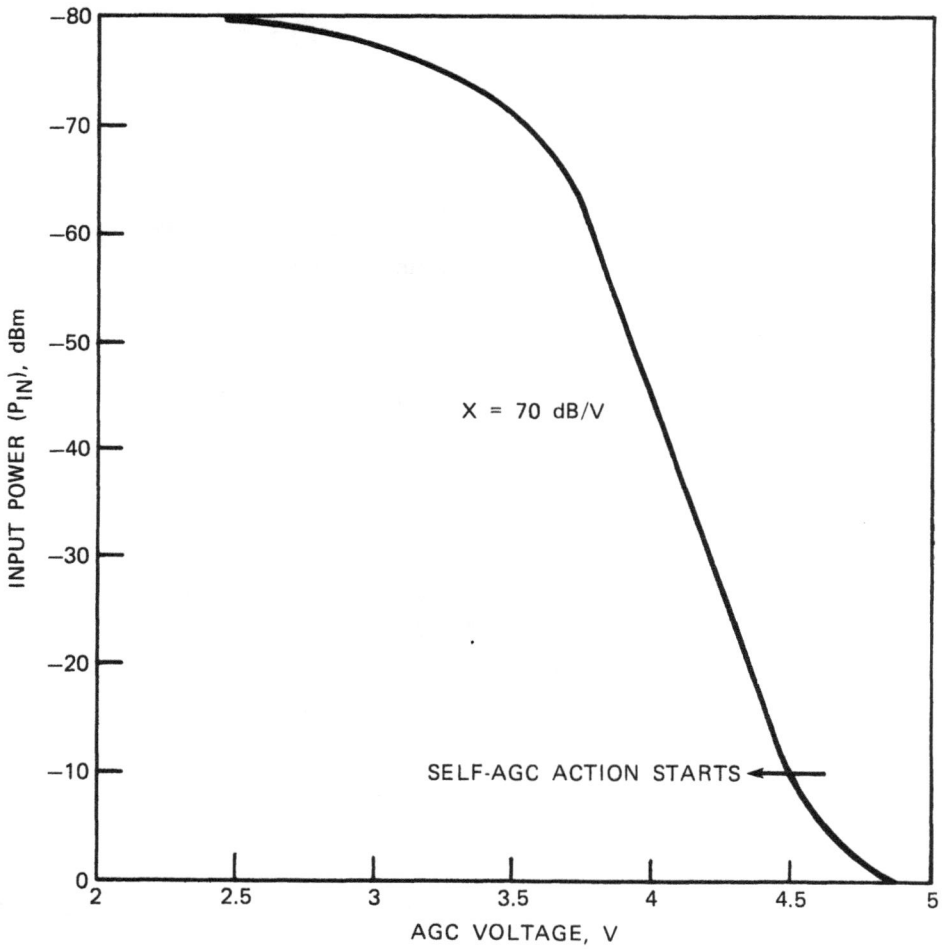

FIGURE 1-42. AGC Voltage Versus Input Power (30-MHz Continuous Wave AGC).

Figure 1-43 illustrates the change in output power versus input power. The output power changes from -14.6 to -15.2 dBm ($P_{in} = -10$ dBm to $P_{in} = -60$ dBm) or ΔP_o (dB) $= 0.6$ dB, which is very close to our design value of 0.5 dB. The loop rise time was measured (with ± 1 dB input steps) and was found to be 65 μs, which is very close to our design value. The measured loop gain is 90, with a predicted value (Figure 1-20) of

$$LG = 0.23 \times 10^{-3} \; K_{SL} \; X \; A_v A_\Delta A_\varepsilon \; 10^{\frac{P_{PD}(dBm)}{10}} \tag{1-118}$$

or

$$LG = (0.23 \times 10^{-3})(500)(70)(392) \; 10^{\frac{-15}{10}} \tag{1-119}$$

or

$$LG = 99.8 \tag{1-120}$$

which is very close to the measured value.

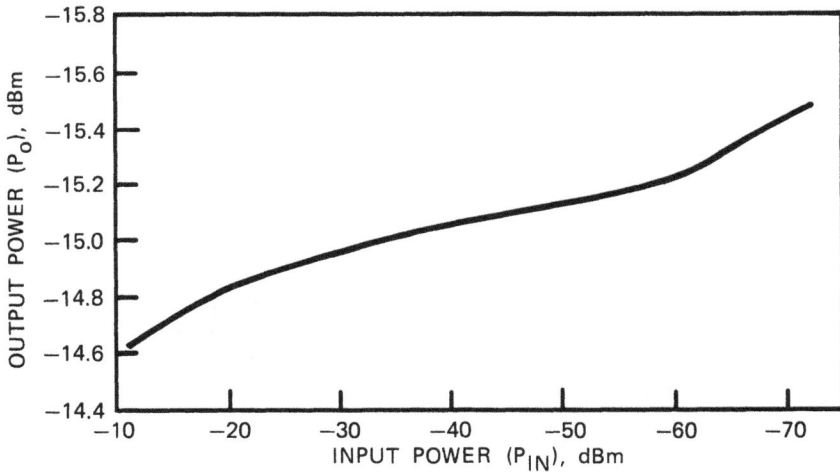

FIGURE 1-43. Output Power Versus Input Power.

This example was chosen not for its state-of-the-art importance, but to illustrate the accuracy afforded by the equations presented. The circuitry used to achieve a given AGC loop is entirely up to the designer.

An AGC loop for a pulsed conical scanning radar receiver will now be presented.

Conical Scanning AGC Loop

The AGC loop illustrated in Figure 1-44 is typical of many conical scanning radar receivers. The AGC loop keeps the video (e_N) constant for slowly varying intensity changes. However, the AGC loop must have minimum effect at the conical scanning frequency to avoid any plane rotation. Figure 1-45a illustrates a typical conical scanning radar. The angle return is dependent on the modulation envelope of the return pulses (Figure 1-45b) and phase, with respect to a reference voltage (Figure 1-45c). The resultant azimuth and elevation signals are defined in Figure 1-45d.

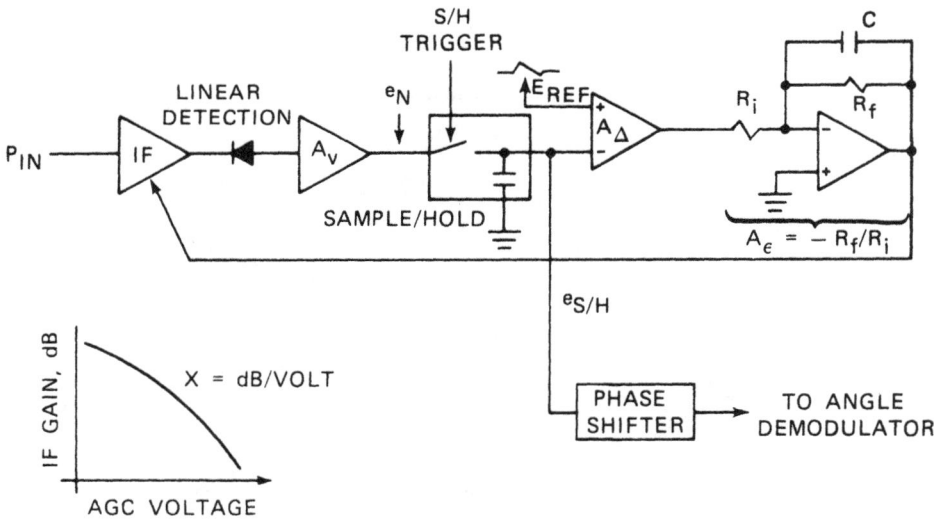

FIGURE 1-44. Conical Scanning Automatic Gain Control Loop.

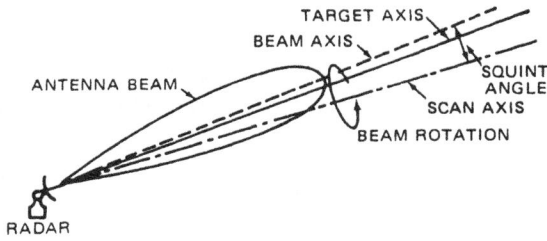

(a) TYPICAL CONICAL SCAN RADAR.

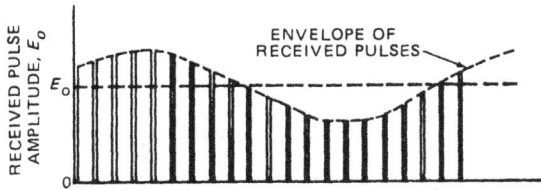

(b) RECEIVED TARGET ENVELOPE.

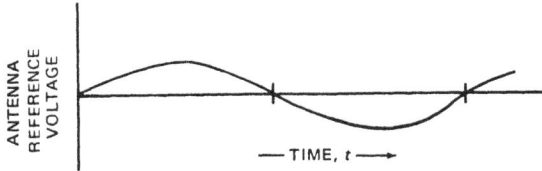

(c) REFERENCE VOLTAGE.

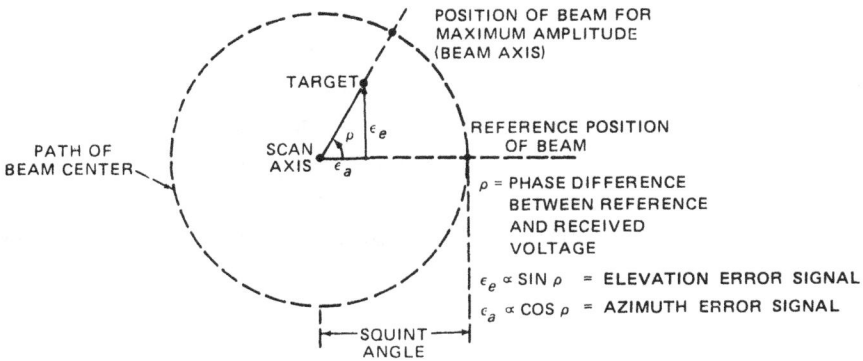

ρ = PHASE DIFFERENCE BETWEEN REFERENCE AND RECEIVED VOLTAGE

$\epsilon_e \propto \text{SIN } \rho$ = ELEVATION ERROR SIGNAL

$\epsilon_a \propto \text{COS } \rho$ = AZIMUTH ERROR SIGNAL

(d) AZIMUTH AND ELEVATION SIGNALS.

FIGURE 1-45. Typical Conical Scanning Radar.

The information necessary for angle tracking is usually obtained by comparing the sampled received signal with a reference signal (Figures 1-45b and 1-45c); thus, the signal modulation envelope should have minimum AGC at the scanning frequency. (If AGC were applied at the scanning frequency, no modulation envelope would result, due to the inherent input modulation reduction characteristics of the AGC loop.) Furthermore, any phase shift of the received modulation envelope must be kept small to avoid cross talk between the azimuth and elevation angle loops (plane rotation). The loop gain at the scanning frequency must be much less than unity.

Assuming a linear detector, the modulation ratio of the normalized video (Me_N) to the intermediate frequency (IF) input (Me_{in}) may be given as [8]:

$$\frac{Me_N}{Me_{in}} = \frac{1}{1 + LGY_{AGC}(f)} \tag{1-121}$$

where

$LG =$ low frequency loop gain (Figure 1-21)

$Y_{AGC}(f)$ = filter and sample hold transfer function at frequency f

Since $Y_{AGC}(f)$ is a complex quantity, Me_N will also be complex. In certain conical scanning radar receivers, the scan modulation is extracted from the sample/hold output, and it is important that the phase shift between the input modulation and output modulation be insensitive to any dynamic receiver parameters, or the coordinates into which the target alignment is resolved will not coincide with the vertical and horizontal coordinates. (Any fixed phase shift can easily be compensated by a phase shift circuit, as illustrated in Figure 1-44.)

The low frequency loop gain for the AGC configuration of Figure 1-44 may be given as (Figure 1-21)

$$LG_{Lin} = 0.115 \times A_\Delta A_\varepsilon e_N \tag{1-122}$$

66

The static regulation for a linear detector may be given as (Figure 1-21)

$$\Delta e_N(dB) = 20 \log \left(\frac{\Delta P_{in}(dB)}{X A_\Delta A_\epsilon e_N} + 1 \right) \qquad (1\text{-}123)$$

where

$\Delta e_N(dB)$ = decibel change in normalized video for a given change in $\Delta P_{in}(dB)$.

Equation (1-123) may be given as

$$\Delta e_N(dB) = 20 \log \left(\frac{\Delta AGC}{A_\Delta A_\epsilon e_N} + 1 \right) \qquad (1\text{-}124)$$

where

$$\Delta AGC = \frac{\Delta P_{in}(dB)}{X} \qquad (1\text{-}125)$$

Equation (1-124) may now be solved in terms of $A_\Delta A_\epsilon e_N$ required for a desired $\Delta e_N(dB)$ and a specified ΔAGC:

$$A_\Delta A_\epsilon e_N = \frac{\Delta AGC}{10^{\left(\frac{\Delta e_N(dB)}{20}\right)} - 1} \qquad (1\text{-}126)$$

Now, since $A_\Delta A_\epsilon e_N$ is known for given static regulation requirements, the loop gain is also uniquely determined from Equation (1-122).

The plane rotation as a function of AGC loop dynamics is fairly complicated. However, provided that the signal PRF is much larger than the conical scanning frequency, the plane rotation may be reasonably approximated as [8]

$$\theta_{PR} \simeq \text{Tan}^{-1} \left(\frac{LG \, f_{3dB}(LPF)}{f_{cs}} \right) \qquad (1\text{-}127)$$

where

θ_{PR} = plane rotation due to AGC loop

f_{cs} = conical scanning frequency

$f_{3dB}(\text{LPF})$ = low pass 3-dB frequency response necessary for given θ_{PR}.

Solving Equation (1-127), f_{3dB} may be given as

$$f_{3dB}(\text{LPF}) = \frac{f_{cs}}{LG} \, \text{Tan} \, \theta_{PR} \qquad (1\text{-}128)$$

A practical design example will now be given.

Design parameters: minimum input power for AGC action = -60 dBm, maximum input for linear operation = -10 dBm, conical scanning frequency = 175 Hz, plane rotation = 5 degrees, $\Delta e_N = \pm 1$ dB, instantaneous input dynamic range = ± 10 dB.

To obtain a ± 10 dB instantaneous input dynamic range, a linear detector will be utilized ($\Delta e_N(\text{dB}) = \Delta P_{in}(\text{dB})$). Figure 1-46 illustrates the linear characteristics for a Schottky diode, and, as can be seen, the detector output is quite linear (dB out versus dB in) for inputs greater than -3 dBm. A normalized value of $P_{PD}(\text{dBm})_N = +7$ will be used ($P_{PD}(\text{max}) = +17$ dBm, $P_{PD}(\text{min}) = -3$ dB). PIN diode attenuators (with a linearizing circuit) will be utilized to obtain the necessary gain variation as illustrated in Figure 1-47.

Figure 1-48 illustrates the basic AGC loop. A normalized video voltage (e_N) of -3 volts will be used. Thus, for a normalized $P_{PD}(\text{dBm}) = +7$ dBm (Figure 1-46)

$$A_v = \frac{3}{0.6} = 5 \qquad (1\text{-}129)$$

$$K_{LIN} = e_D(mV)\ 10^{\frac{-P_{PD}(dBm)}{20}}$$

$$K_{LIN} = (460\ mV)\ 10^{\frac{-5}{20}} = 258\ mV/mW$$

$e_D = 460\ mV$

$P_{PD} = +5$

DETECTOR OUTPUT VOLTAGE, V

INPUT POWER, dBm

HP 5082-2800

+15 V

100 kΩ

100 pF

0.1 μF

$P_{PD}(dBm)$

e_D

100 Ω

27 pF

2K

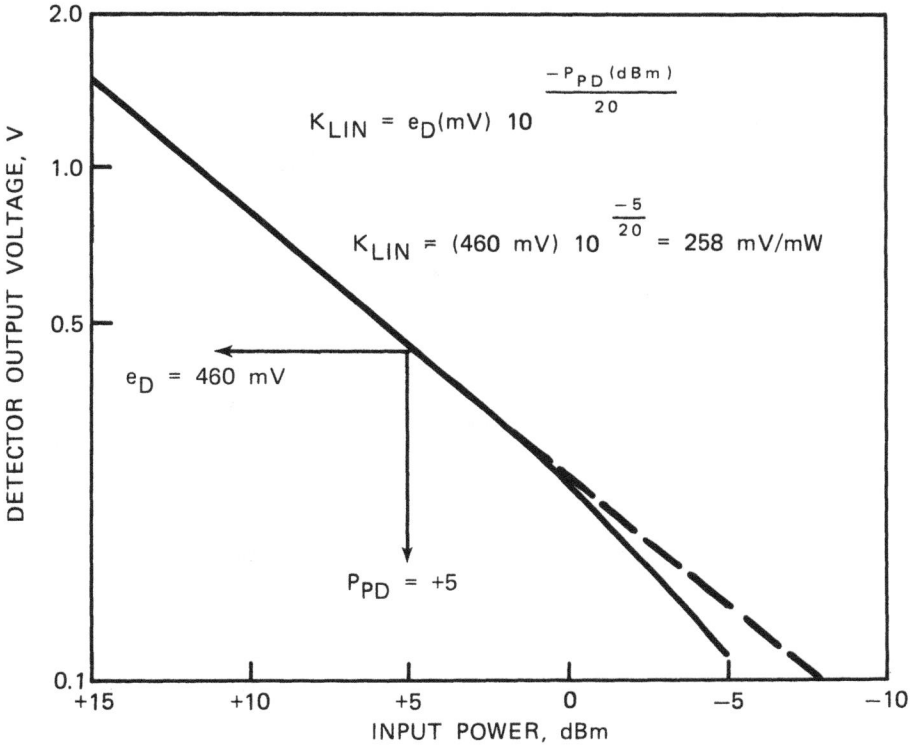

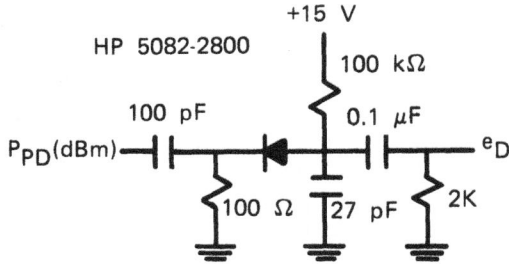

FIGURE 1-46. Linear Detector Characteristics.

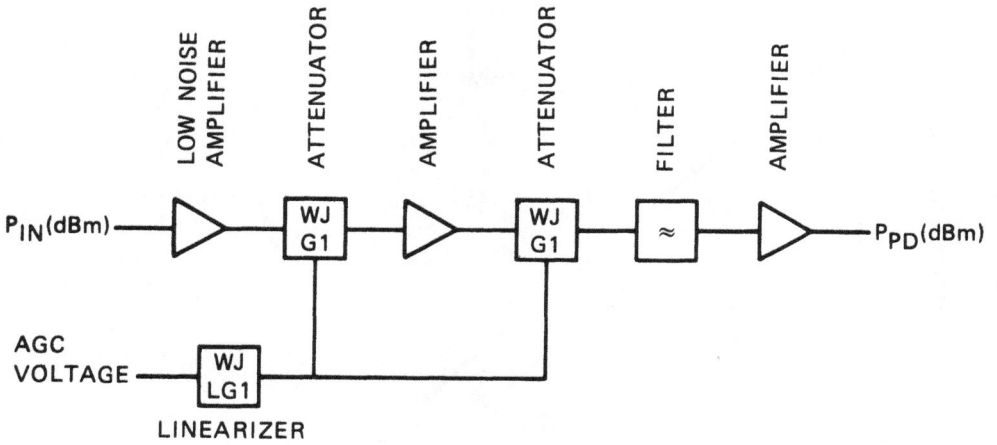

FIGURE 1-47. Basic Conical Scanning Receiver.

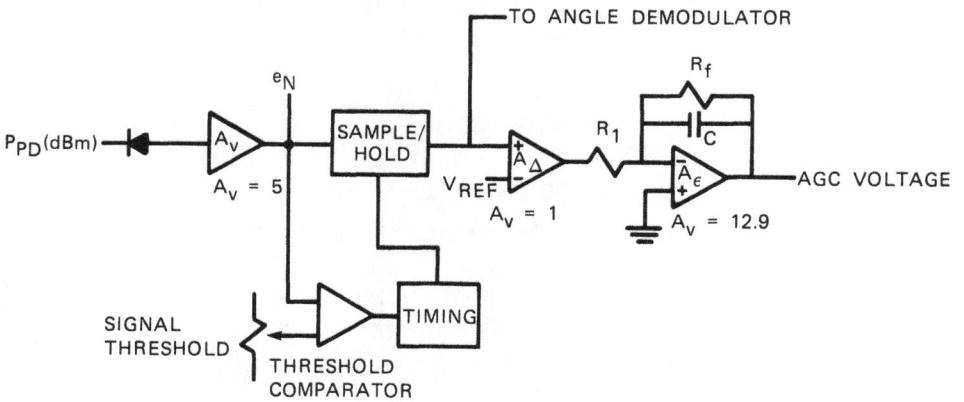

FIGURE 1-48. Basic Conical Scanning AGC Loop.

70

Figure 1-49 illustrates the AGC characteristics for the receiver ($A_V = 5$ and $e_N = 3$ volts). The variable gain slope, X, is 5 dB/volt, and $A_\Delta A_\varepsilon e_N$ (Equation (1-126)) may now be found

$$A_\Delta A_\varepsilon e_N = \frac{10}{10^{\frac{2}{20}} - 1} = 38.6 \tag{1-130}$$

and A_ε is (for $A_\Delta = 1$)

$$A_\varepsilon = \frac{38.6}{(1)(3)} = 12.9 \tag{1-131}$$

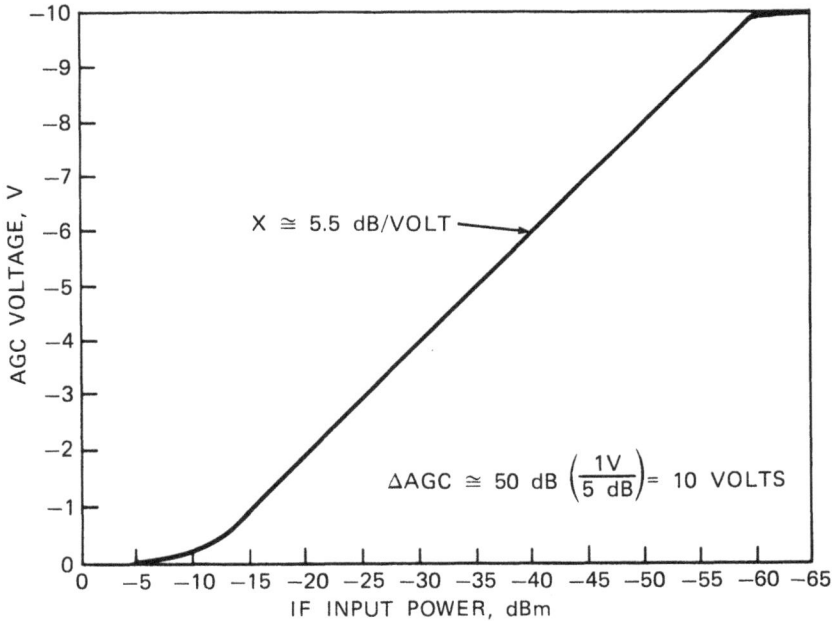

FIGURE 1-49. AGC Voltage Versus Input Power.
$e_N|_{\text{video}} \simeq 3$ V at -30 dBm.

The loop gain necessary for our static regulation is (Equation (1-122))

$$LG = (0.115)(5)(38.6) = 22.2 \tag{1-132}$$

The low-pass filter 3-dB frequency response ($f_{3dB}(LPF)$) necessary to ensure the loop gain is low enough for the desired 5-degree plane rotation is (Equation (1-128))

$$f_{3dB}(LPF) = \frac{175}{22.2} \, Tan \, 5 = 0.69 \, Hz \tag{1-133}$$

The values for R_f, C_f and R_1 (Figure 1-48) will now be found

$$f_{3dB}(LPF) = \frac{1}{2\pi R_f C_f} \tag{1-134}$$

or

$$R_f C_f = \frac{0.159}{f_{3dB}(LPF)} = 0.23 \tag{1-135}$$

Letting $C = 6.6 \, \mu F$,

$$R_f = \frac{0.23}{6.6 \times 10^{-6}} \approx 34.8 \, k\Omega \tag{1-136}$$

and R_1 (for $A_\varepsilon = 12.9$) is

$$R_1 = \frac{R_f}{A_\varepsilon} = \frac{34.8}{12.9} \approx 2.71 \, k\Omega \tag{1-137}$$

Figure 1-50 illustrates the receiver and AGC loop characteristics. As can be seen, $\Delta e_N(\text{dB})$ is quite close to our design value (± 1 dB). The sample/hold signal is held for the pulse repetition interval, thus the duty cycle (D) is unity, and the predicted rise time (Figure 1-21) is

$$\iota_r = \frac{2.2\text{RC}}{\text{LG}} = \frac{2.2(34.8 \times 10^3)(6.6 \times 10^{-6})}{22.1} \qquad (1\text{-}138)$$

or

$$\iota_r \simeq 23 \text{ msec} \qquad (1\text{-}139)$$

which compares favorably with the measured value of 27 msec.

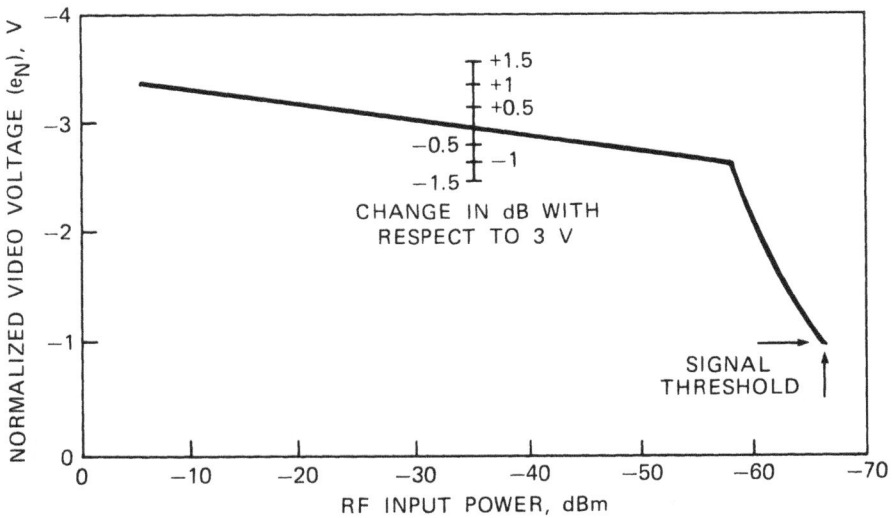

FIGURE 1-50. Receiver/AGC Variable Gain Characteristics.

The sample/hold process complicates the loop stability, and the loop will oscillate half the sampling frequency (PRF) if the loop gain at this frequency exceeds unity [8].

Optimizing Loop Rise Times

The loop rise time calculations presented are valid only for small steps in input power, typically less than ± 2 dB. Tables 1-2 and 1-5 illustrate the dependence of the input step and polarity on the loop rise time. As can be seen, the rise time decreases for positive input steps (increased power), and increases for negative steps (decreased power). This nonlinearity is due to the inherent nonlinear properties of the detector. Figure 1-51 illustrates the normalized instantaneous detector output for a square law detector, using

$$e_{D,SL} = 10^{\frac{\Delta P_{PD}(dBm)}{10}} \qquad (1\text{-}140)$$

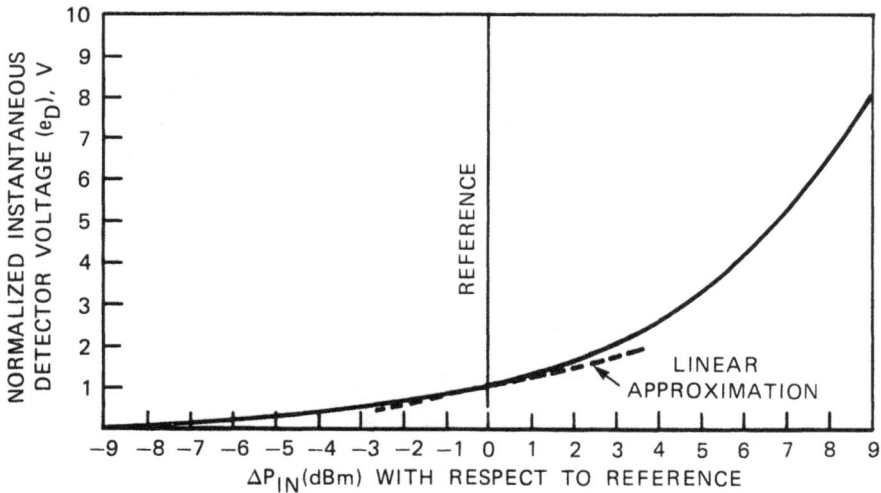

FIGURE 1-51. Instantaneous Detector Voltage Versus $\Delta P_{in}(dBm)$.

As can be seen, e_D is very nonlinear for ΔP_{in}(dB) greater than ± 2 dB. Thus the AGC integrator, or low-pass-filter input, will be large for $+\Delta P_{in}$(dB) and small for $-\Delta P_{in}$(dB) (e.g., for ΔP_{in}(dB) $= +5$, $\Delta e_D \simeq 3.2 - 1 = 2.2$, and for ΔP_{in}(dB) $= -5$, $\Delta e_D = 1 - 0.3 = 0.7$). Thus the rise times for large values of $+\Delta P_{in}$(dB) should be expected to be shorter than for large values of $-\Delta P_{in}$(dB).

Certain AGC loops may require a more constant loop rise time with large values of $\pm \Delta P_{in}$(dB). Two solutions to this problem are apparent from the previous discussion: (1) have a shorter RC time constant for $-\Delta P_{in}$(dB) steps, or (2) linearize e_D versus ΔP_{in}(dB). These two techniques will now be presented.

Variable Time Constant

Figure 1-52 illustrates a variable time constant circuit. This circuit replaces the differencing circuit and integrator of Figure 1-9. The diode, D_1, blocks positive voltages (or $+\Delta P_{in}$(dB)), and the loop behaves as it normally does; however, D_1 turns on for $-\Delta P_{in}$(dB), thus decreasing the integrator's RC time constant (R_1 and R_2 are now in parallel). The transistor compensates for the 0.6-volt DC diode drop. R1 may now be adjusted to give equal rise times for $+\Delta P_{in}$(dB). This circuit works for ΔP_{in}(dB) $\simeq \pm 5$ dBm. A much better solution would be to linearize e_N with respect to ΔP_{in}(dB).

Linearized Time Constant

A logarithmic amplifier preceding the differencing amplifier (Figure 1-53) will linearize e_N with respect to ΔP_{in}(dBm). A logarithmic amplifier has the following characteristics [9]

$$e_N = K_1 \log K_2 (A_v e_D) \qquad (1\text{-}141)$$

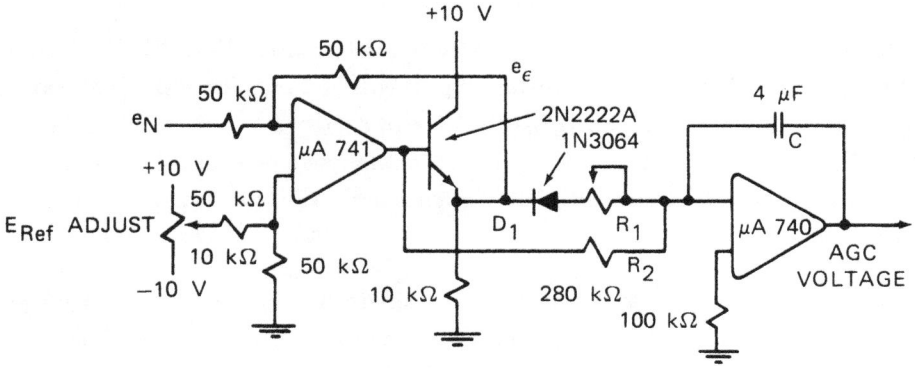

FIGURE 1-52. Variable Time Constant Circuit.

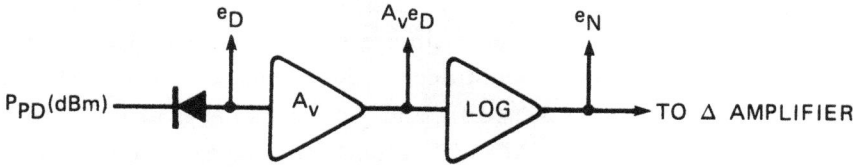

FIGURE 1-53. AGC Loop Employing Logarithmic Amplifier.

where

K_1, K_2 = logarithmic amplifier constants

Now , for a square law detector,

$$e_D = K_{SL} \; 10^{\dfrac{P_{PD}(dBm)}{10}} \tag{1-142}$$

76

and substituting Equation (1-142) into (1-141),

$$e_N = K_1 \log K_2 A_v \left[K_{SL} \, 10^{\frac{P_{PD}(dBm)}{10}} \right] \tag{1-143}$$

or

$$e_N = \frac{K_1 P_{PD}(dBm)}{10} = K_1 \log (K_2 A_v K_{SL}) \tag{1-144}$$

Taking the derivative of Equation (1-144) with respect to P_{PD}(dBm),

$$\frac{\Delta e_N}{\Delta P_{PD}(dB)} = \frac{K_1}{10} \quad (V/dB) \tag{1-145}$$

or, the logarithmic video output has a constant slope of $K_1/10$(V/dB), which is the desired result.

To calculate the static regulation, loop gain, and loop rise time, simply include the logarithmic amplifier in the calculations given in Appendices C, D, and E. Figures 1-54 and 1-55 illustrate the pertinent parameters for the logarithmic amplifier AGC loop.

Figure 1-56 illustrates a basic pulse AGC loop incorporating a logarithmic amplifier for optimizing loop rise times.

77

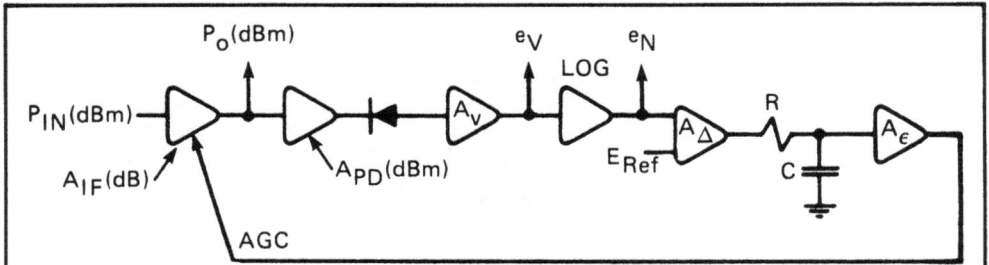

X = IF GAIN SLOPE (dB/V)   LS = LOGARITHMIC SLOPE (V/dBm)
                                 D = UPDATE DUTY CYCLE

STATIC REGULATION

$$\Delta P_{o,SL}\,(dBm) = \frac{\Delta AGC}{2A_\Delta A_\epsilon (LS)} \qquad \Delta e_{N,SL}\,(dBm) = \frac{\Delta AGC}{A_\Delta A_\epsilon (LS)}$$

LOOP RISE TIME (10 TO 90%)

$$\tau_{r,SL} = \frac{1.05\,RC}{X A_\Delta A_\epsilon (LS) D}$$

LOOP GAIN

$$LG_{SL} = 2\,X A_\Delta A_\epsilon (LS)$$

$$f_{3dBV}(LG) = \frac{0.159}{(LG)\,RC} \qquad \tau_r = \frac{2.2\,RC}{D(LG)}$$

FIGURE 1-54. Low-Pass Filter and Log Video AGC Design Summary.

78

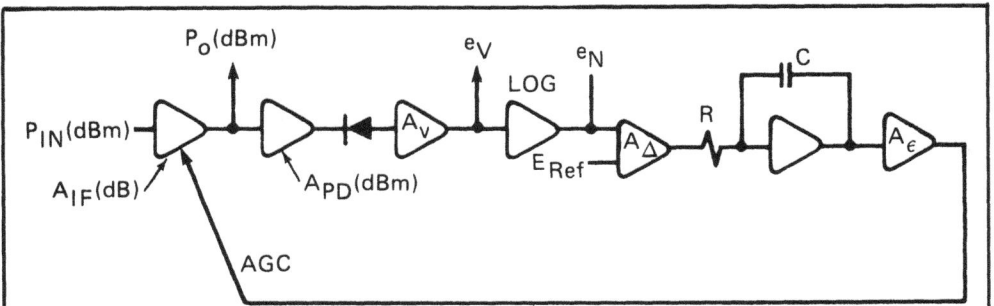

FIGURE 1-55. Integrator and Log Video AGC Design Summary.

X = IF GAIN SLOPE (dB/V)   LS = LOGARITHMIC SLOPE (dB/V)
                            D = INTEGRATOR UPDATE DUTY
                                CYCLE

STATIC REGULATION

$$\Delta P_o(dBm) \simeq 0 \qquad\qquad \Delta e_N(dBm) \simeq 0$$

LOOP RISE TIME (10 TO 90%)

$$t_{r,SL} = \frac{1.05\,RC}{X A_\Delta A_\varepsilon (LS) D} \qquad\qquad t_{r,Lin} = \frac{2.1\,RC}{X A_\Delta A_\varepsilon (LS) D}$$

LOOP GAIN

$$LG_{SL} = \frac{0.315\, X A_\Delta A_\varepsilon (LS) D}{fRC} \qquad\qquad LG_{Lin} = \frac{0.158\, X A_\Delta A_\varepsilon (LS) D}{fRC}$$

$$(LG)(t_r) = \frac{0.331}{f}$$

f = INPUT MODULATION FREQUENCY

79

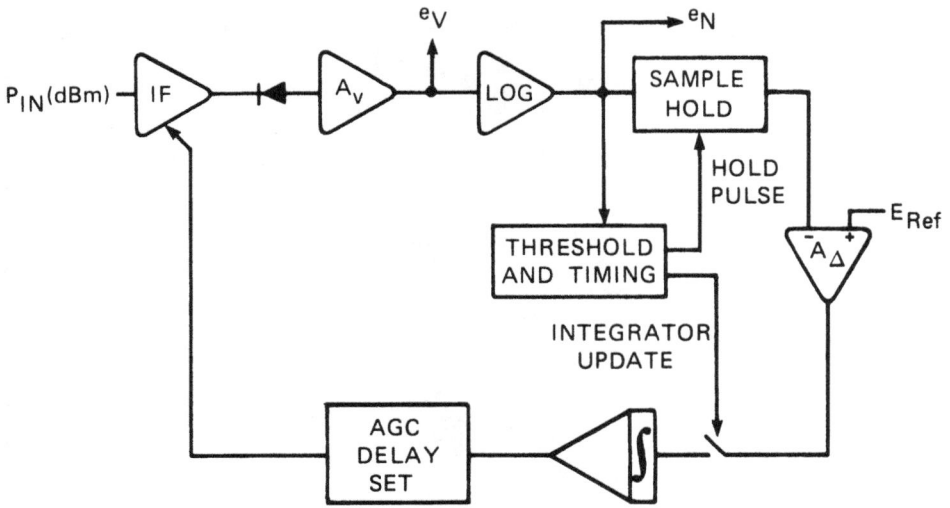

FIGURE 1-56. Linear Rise Time AGC Loop.

Basic Loop Stability

AGC loop dynamics have been given as (Figures 1-5 and 1-6 and Equation (1-13))

$$M_o = \frac{M_I}{1 + LG(f)} \tag{1-146}$$

where

M_o = output modulation

M_I = input modulation

$LG(f)$ = frequency dependent loop gain

Classical feedback theory tells us that the loop gain must be less than unity when its phase shift is 180 degrees. The AGC loop designer must ensure that this condition is satisfied. The loop frequency response is usually dominated by the low-pass filter or integrator; thus, to avoid instability, the variable gain element frequency response must be much larger than: (1) f_{3dB}(LPF) for the low-pass filter and (2) the frequency for unity loop gain for the integrator.

References

1. Oliver, B. M. "Automatic Volume Control as a Feedback Problem," in *Proceedings of the IRE*, April 1948, pp. 466-473.

2. Hughes, R. S. "Diodes Make Good Gain-Control Devices," *Electronic Design*, 3 February 1970.

3. Viles, R. S. "Need a PIN-Diode Attenuator?" *Electronic Design*, 29 March 1977, pp. 100-102.

4. Hewlett Packard. *High Performance PIN Attenuators for Low Cost Applications*. Palo Alto, CA. (Application Note 936, publication UNCLASSIFIED.)

5. Hewlett Packard. *An Attenuator Design Using PIN Diodes*. Palo Alto, CA. (Application Note 9R, publication UNCLASSIFIED.)

6. Fisher, D. and Kritger, T. "A Linear GaAs MMIC Variable Attenuator," *RF Design*, October 1977, pp. 27-31.

7. Shurmer, H. V. *Microwave Semiconductor Devices*. John Wiley and Sons, New York, 1971.

8. Naval Weapons Center. *Effect of Automatic Gain Control and Sample-Hold Characteristics on Plane Rotation of Certain Conical Scan Radar Receivers*, by Richard Smith Hughes. China Lake, CA, NWC, August 1983. (NWC TP 6454, publication UNCLASSIFIED.)

9. Hughes, R. S. *Logarithmic Amplification with Application to Radar and EW*. Artech House, Inc., Dedham, MA, 1986.

10. Watson, H. A. *Microwave Semiconductor Devices and Their Circuit Applications*. McGraw-Hill Book Company, New York, 1969.

Appendix 1A

POWER-VOLTAGE RELATIONSHIPS
FOR A 50-OHM SYSTEM

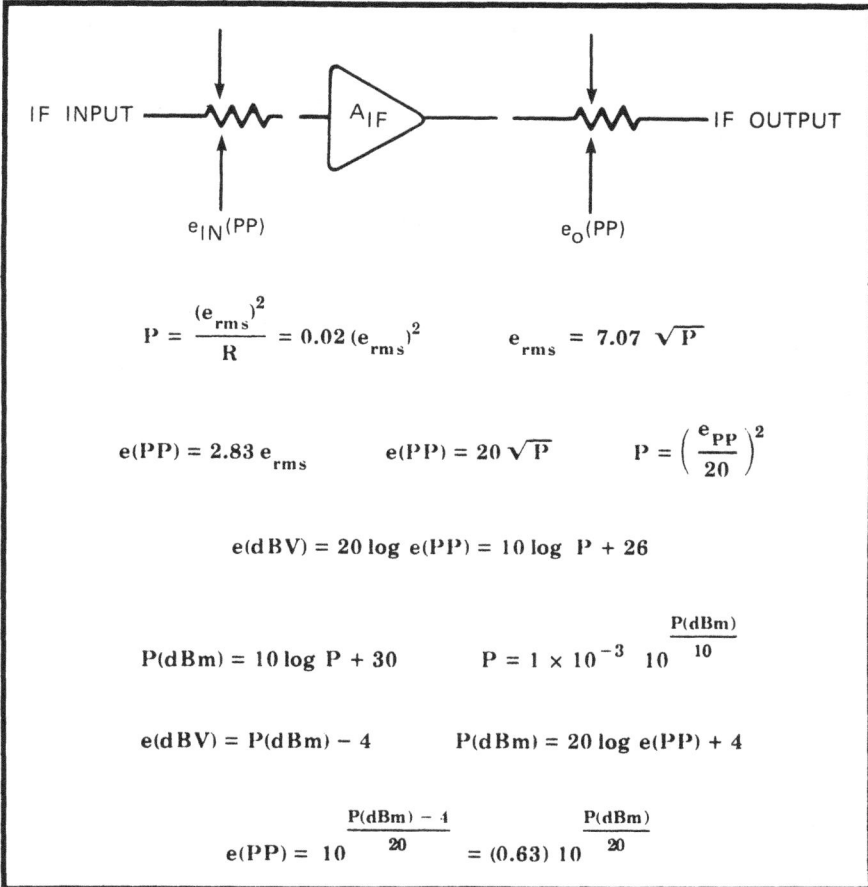

$$P = \frac{(e_{rms})^2}{R} = 0.02\,(e_{rms})^2 \qquad e_{rms} = 7.07\,\sqrt{P}$$

$$e(PP) = 2.83\,e_{rms} \qquad e(PP) = 20\,\sqrt{P} \qquad P = \left(\frac{e_{PP}}{20}\right)^2$$

$$e(dBV) = 20\log e(PP) = 10\log P + 26$$

$$P(dBm) = 10\log P + 30 \qquad P = 1 \times 10^{-3}\;10^{\frac{P(dBm)}{10}}$$

$$e(dBV) = P(dBm) - 4 \qquad P(dBm) = 20\log e(PP) + 4$$

$$e(PP) = 10^{\frac{P(dBm)-4}{20}} = (0.63)\,10^{\frac{P(dBm)}{20}}$$

FIGURE 1A-1. Power-Voltage Relationships
for a 50-Ohm System.

83

Appendix 1B

BASIC DETECTOR CHARACTERISTICS

The crystal diode detector converts the IF waveform into a dc waveform for continuous wave AGC, or into pulses for pulse AGC. The basic characteristics for detectors are square law for low input powers (the output voltage increases 2 dBV for each dBm increase in input power) and linear for high input powers (the output voltage increases 1 dBV for each dBm increase in input power). Watson [10] covers detector characteristics in detail; however, several important characteristics pertinent to AGC design will be presented here.

Figure 1B-1 illustrates the input-output characteristics for a typical Schottky barrier detector (HP 5082-2800). The square law characteristics will be covered first, then the linear.

The square law detector characteristics are valid for inputs of -15 dBm and lower. The detector output voltage, e_D, may be given as

$$e_D = K_{SL} P_{PD} \qquad (1B\text{-}1)$$

where

e_D is in millivolts

P_{PD} is in milliwatts

K_{SL} is a square law detector constant

The diode constant, K_{SL}, is the detector static gain,

$$K_{SL} = \frac{e_D}{P_{PD}} \left(\frac{mV}{mW} \right) \qquad (1B\text{-}2)$$

84

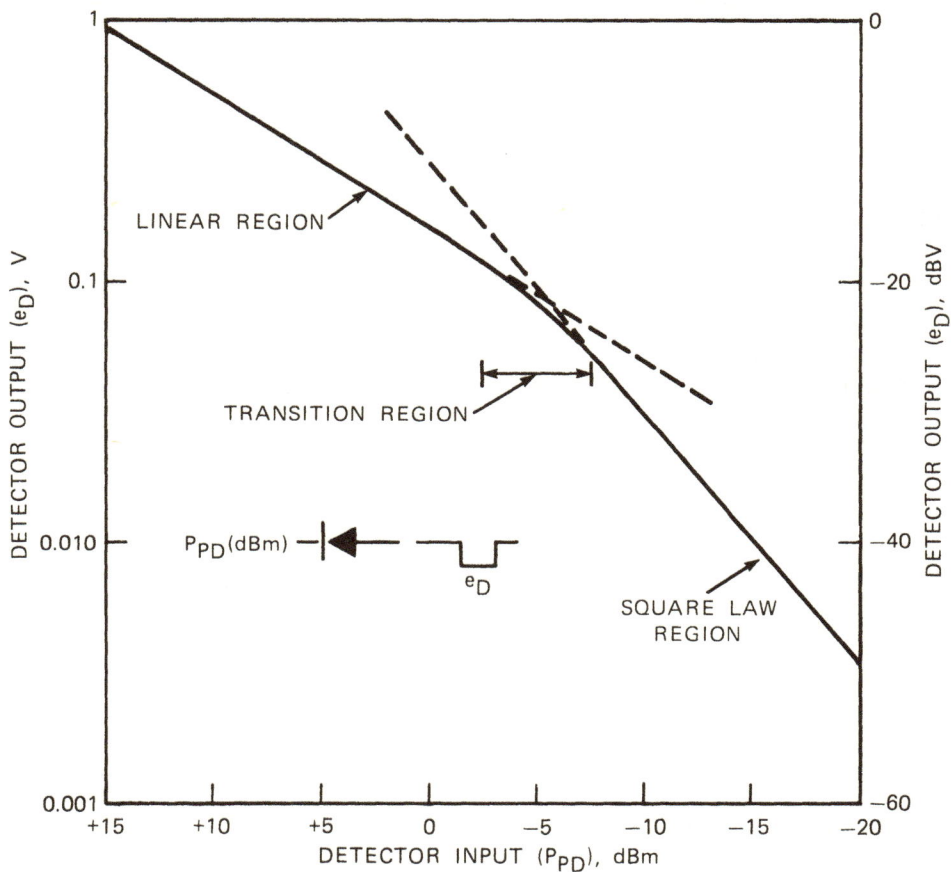

FIGURE 1B-1. Detector Characteristics (see Figure 1-16).

and is fairly constant for detector inputs less than -15 dBm. The value of K_{SL} may be found, provided a curve similar to Figure 1B-1 is given.

The detector input power, P_{PD}, may be given in terms of dBm as (Appendix 1A)

$$P_{PD}(mW) = 10^{\frac{P_{PD}(dBm)}{10}} \qquad (1B\text{-}3)$$

substituting Equation (1B-3) into (1B-2),

$$K_{SL} = e_D \ 10^{\frac{-P_{PD}(dBm)}{10}} \ \left(\frac{mV}{mW} \right) \qquad (1B\text{-}4)$$

Referring to Figure 1B-1, K_{SL} is (for $P_{PD} = -20$ dBm, $e_D = 3.5$ mV)

$$K_{SL} = (3.5) \ 10^{\frac{20}{10}} \ \left(\frac{mV}{mW} \right) \qquad (1B\text{-}5)$$

or

$$K_{SL} = 350 \ \left(\frac{mV}{mW} \right) \qquad (1B\text{-}6)$$

Knowing K_{SL}, e_D for a $P_{PD}(dBm)$ may be found by solving Equation (1B-4) for e_D:

$$e_{D,SL} = K_{SL} \ 10^{\frac{P_{PD}(dBm)}{10}} \ \left(mV \right) \qquad (1B\text{-}7)$$

The dynamic detector gain, $A_{D,SL}$ (V/V), is necessary in calculating the loop gain (Figure 1-6). The detector's dynamic gain will be found for a fixed input, $e_{PD}(PP)$:

$$A_{D,SL} \ (V/V) = \frac{d \ e_D}{d \ e_{PD}(PP)} \ \left(e_{PD}(PP) \right) = \text{Constant} \qquad (1B\text{-}8)$$

86

where (Appendix 1A)

$$e_{PD}(PP) = (0.63) \ 10^{\frac{P_{PD}(dBm)}{20}} \tag{1B-9}$$

or

$$e_{PD}(PP) = (0.63) \ 10^{\left(\frac{P_{PD}(dBm)}{10}\right)^{1/2}} \tag{1B-10}$$

and

$$10^{\frac{P_{PD}(dBm)}{10}} = \left(\frac{e_{PD}(PP)}{0.63}\right)^{1/2} \tag{1B-11}$$

Thus,

$$10^{\frac{P_{PD}(dBm)}{10}} = 2.51 \left[e_{PD}(PP)\right]^2 \tag{1B-12}$$

Substituting Equation (1B-12) into (1B-7),

$$e_D = \left(2.51\right)\left(K_{SL}\right) \left[e_{PD}(PP)\right]^2 \ \ (mV) \tag{1B-13}$$

The dynamic detector gain, $A_{D,SL}$ (V/V), is

$$A_{D,SL}(V/V) = \frac{d \ e_D}{d \ e_{PD}(PP)} \tag{1B-14}$$

Thus from Equation (1B-13),*

$$A_{D,SL}(V/V) \simeq 5 \times 10^{-3} \ K_{SL} \ e_{PD} \ (PP) \tag{1B-15}$$

---

*K_{SL} has been defined in terms of e_D in millivolts ($K_{SL} = e_D(mV)/P_{PD}(mV)$). Thus K_{SL} must be divided by 1,000 for e_D in volts.

87

Since (Appendix 1A)

$$e_{PD}(PP) = (0.63) \; 10^{\frac{P_{PD}(dBm)}{20}} \tag{1B-16}$$

$$A_{D,SL}(V/V) = 3.76 \times 10^{-3} \; K_{SL} \; 10^{\frac{P_{PD}(dBm)}{20}} \tag{1B-17}$$

The dynamic detector gain in V/dBm, $A_{D,SL}$(V/dBm), is needed when solving for the loop rise time, τ_r. This gain may be found by differentiating Equation (1B-7). The solution to this differentiation is

$$A_{D,SL}(V/dBm) = 0.23 \times 10^{-3} \; K_{SL} \; 10^{\frac{P_{PD}(dBm)}{10}} \tag{1B-18}$$

The voltage gains for the diode illustrated in Figure 1B-1 are, assuming $P_{PD}(dBm) = -30$ dBm and $K_{SL} = 350$,

$$A_{D,SL}(V/V) = (3.16 \times 10^{-3})(350) \; 10^{\frac{-30}{20}} = 0.035 \; V/V \tag{1B-19}$$

$$A_{D,SL}(V/dBm) = (0.23 \times 10^{-3})(350) \; 10^{\frac{-30}{10}} = 80.5 \times 10^{-6} \; V/dBm \tag{1B-20}$$

Figure 1B-2 summarizes square law detector characteristics.

$$P_{PD}(dBm) \!-\!\!\blacktriangleleft\!\!\!-\! e_D(mV)$$

$$K_{SL} = \frac{e_D}{P_{PD}} \left(\frac{mV}{mW} \right)$$

$$e_D = K_{SL} \left(10^{\frac{P_{PD}(dBm)}{10}} \right) \times 10^{-3} \ (V)$$

$$K_{SL} = e_D(mV) \, 10^{\frac{-P_{PD}(dBm)}{10}} \ (mV/mW)$$

$$A_{D,SL} \, (V/V) = 3.16 \times 10^{-3} \, K_{SL} \, 10^{\frac{P_{PD}(dBm)}{20}} = 5 \times 10^{-3} \, K_{SL} \, e_{PD}(pp)$$

$$A_{D,SL} \, (V/dBm) = 0.23 \times 10^{-3} \, K_{SL} \, 10^{\frac{P_{PD}(dBm)}{10}}$$

FIGURE 1B-2. Square Law Detector Summary.

Linear detector characteristics extend for input powers larger than -2 dBm for the HP 5082-2800 detector (Figure 1B-1). The output voltage may be given as

$$e_{D,Lin} = K_{Lin} \sqrt{P_{PD}} \tag{1B-21}$$

where

e_{DLin} is in millivolts

P_{PD} is in milliwatts

K_{Lin} is a linear detector constant

89

Solving Equation (1B-21) for K_{Lin},

$$K_{Lin} = \frac{e_{D,Lin}}{\sqrt{P_{PD}}} \left(\frac{mV}{mW} \right) \tag{1B-22}$$

P_{PD} (mW) may be given as (Appendix 1A)

$$P_{PD}(mW) = 10^{\frac{P_{PD}(dBm)}{10}} \tag{1B-23}$$

Substituting Equation (1B-23) into (1B-21)

$$K_{Lin} = e_{D,Lin}(mV) \, 10^{\frac{-P_{PD}(dBm)}{20}} \tag{1B-24}$$

Referring to Figure 1B-1, for $P_{PD} = +10$ dBm and $e_D = 520$ mV, K_{Lin} is

$$K_{Lin} = (520) \, 10^{\frac{-10}{20}} \tag{1B-25}$$

or

$$K_{Lin} = 164 \tag{1B-26}$$

The dynamic detector gain, $A_{D,Lin}$ (V/V), will be found in terms of e_{PD}(PP)

$$A_{D,Lin}(V/V) = \frac{d \, e_{D,Lin}}{d \, e_{PD}(PP)} \tag{1B-27}$$

Using Appendix 1A

$$e_{PD}(PP) = (0.63) 10^{\frac{P_{PD}(dBm)}{20}} \tag{1B-28}$$

and substituting Equation (1B-28) into (1B-24), and solving for e_D,

$$e_D = 1.59 \, K_{Lin} \, e_{PD} \, (PP) \, (mV) \tag{1B-29}$$

90

The dynamic linear detector gain may now be found (remember K_{Lin} is defined in terms of e_D in mV)

$$A_{D,Lin}(V/V) = 1.59 \times 10^{-3} K_{Lin} \qquad (1B\text{-}30)$$

Thus, for the detector characteristics illustrated in Figure 1B-1 ($P_{PD} = +10\ dBm$, $e_D = 520\ mV$, $K_{Lin} = 164$)

$$A_{D,Lin}(V/V) = 1.59 \times 10^{-3} (164) \qquad (1B\text{-}31)$$

or

$$A_{D,Lin}(V/V) = 0.26 \qquad (1B\text{-}32)$$

The dynamic linear detector gain in V/dBm, $A_{D,Lin}$ (V/dBm), can be found by solving Equation (1B-24) for e_D and then differentiating.

$$e_D = K_{Lin}\, 10^{\frac{P_{PD}(dBm)}{20}} \quad (mV) \qquad (1B\text{-}33)$$

$$A_{D,Lin}(V/dBm) = \frac{d\, e_P}{d\, P_{PD}(dBm)} \qquad (1B\text{-}34)$$

Thus,

$$A_{D,Lin}(V/dBm) = 116 \times 10^{-6}\, K_{Lin} 10^{\frac{P_{PD}(dBm)}{20}} \qquad (1B\text{-}35)$$

and for our linear detector ($P_{PD} = +10\ dBm$),

$$A_{D,Lin}(V/dBm) = (116 \times 10^{-6})(164)\, 10^{\frac{10}{20}} \qquad (1B\text{-}36)$$

or

$$A_{D,Lin}(V/dBm) = 60.16 \times 10^{-3} \qquad (1B\text{-}37)$$

Figure 1B-3 summarizes linear detector characteristics.

$$P_{PD}(dBm) \longrightarrow | \blacktriangleleft \longrightarrow e_D \ (mV)$$

$$K_{LIN} = \frac{e_P}{\sqrt{P}} \left(\frac{mV}{mW} \right)$$

$$e_D = K_{Lin} \left(10^{\frac{P_{PD}(dBm)}{20}} \right) \times 10^{-3} \ (V)$$

$$K_{Lin} = e_D \ (mV) \, 10^{\frac{-P_{PD}(dBm)}{20}} \left(\frac{mV}{mW} \right)$$

$$A_{D,Lin} \ (V/V) = 1.59 \times 10^{-3} \ K_{Lin}$$

$$A_{D,Lin} \ (V/dBm) = 116 \times 10^{-6} \ K_{Lin} \ 10^{\frac{P_{PD}(dBm)}{20}}$$

FIGURE 1B-3. Linear Detector Summary.

Appendix 1C

STATIC REGULATION CALCULATIONS

The static regulation of AGC loops will now be given. Figure 1-2a illustrates the basic loop configuration.

The AGC voltage is

$$AGC = A_\Delta A_\varepsilon (E_{Ref} - A_v e_D) \tag{1C-1}$$

The voltage will vary as the input power is changed. The change in AGC voltage may be given as

$$\Delta AGC = A_\Delta A_\varepsilon A_v \Delta e_D \tag{1C-2}$$

where Δe_D is the maximum permissible change in detector output for the desired change in input power. Δe_D may be given as

$$\Delta e_D = e_{D,max} - e_{D,min} \tag{1C-3}$$

e_D is related to detector input power, for a square law detector, by (Appendix 1B)

$$e_D = K_{SL} \, 10^{\frac{P_{PD}(dBm)}{10}} \quad (mV) \tag{1C-4}$$

Substituting Equation (1C-4) into (1C-3),

$$\Delta e_D = K_{SL} \left[10^{\frac{P_{PD,max}(dBm)}{10}} - 10^{\frac{P_{PD,min}(dBm)}{10}} \right] \tag{1C-5}$$

93

This equation may be rewritten as

$$\Delta e_D = K_{SL} \, 10^{\dfrac{P_{PD,min}(dBm)}{10}} \left[\dfrac{10^{\dfrac{P_{PD,max}(dBm)}{10}}}{10^{P_{PD,min}(dBm)}} - 1 \right] \qquad (1C\text{-}6)$$

or

$$\Delta e_D = e_{D,min} \left[10^{\dfrac{\Delta P_{PD}(dBm)}{10}} - 1 \right] \qquad (1C\text{-}7)$$

where

$$\Delta P_{PD,SL}(dBm) = P_{PD,max}(dBm) - P_{PD,min}(dBm)$$

Substituting Equation (1C-7) into (1C-2) yields

$$\Delta AGC = A_v A_\Delta A_\varepsilon e_{D,min} \left[10^{\dfrac{\Delta P_{PD,SL}(dBm)}{10}} - 1 \right] \qquad (1C\text{-}8)$$

and solving Equation (1C-8) for $\Delta P_{PD}(dBm)$ gives

$$\Delta P_{PD,SL}(dBm) = 10 \, \log \left(\dfrac{\Delta AGC}{A_v A_\Delta A_\varepsilon e_{D,min}} + 1 \right) \qquad (1C\text{-}9)$$

Since the variable gain IF and predetector IF amplifiers are linear,

$$\Delta P_{PD,SL}(dBm) = \Delta P_{IF}(dBm) \qquad (1C\text{-}10)$$

and

$$\Delta P_{IF,SL}(dBm) = 10 \, \log \left(\dfrac{\Delta AGC}{A_v A_\Delta A_\varepsilon e_{D,min}} + 1 \right) \qquad (1C\text{-}11)$$

94

The average value for $e_{D,N}$ is

$$e_{D,N} = \frac{e_{D,max} + e_{D,min}}{2} \qquad \text{(1C-12)}$$

If $e_{D,N}$ is adjusted in the mid-AGC range, $\Delta AGC/2$,*

$$\Delta P_{IF,SL}(dBm) = 10 \log\left(\frac{\Delta AGC}{A_v A_\Delta A_\varepsilon e_{D,N}} + 1\right) \qquad \text{(1C-13)}$$

The same procedure can be used to find the static regulation for linear detectors:

$$\Delta P_{IF,Lin}(dBm) = 20 \log\left(\frac{\Delta AGC}{A_v A_\Delta A_\varepsilon e_{D,N}} + 1\right) \qquad \text{(1C-14)}$$

Noting that

$$A_v e_{D,N} = e_N \qquad \text{(1C-15)}$$

Equations (1C-13) and (1C-14) become

$$\Delta P_{IF,SL}(dBm) = 10 \log\left(\frac{\Delta AGC}{A_\Delta A_\varepsilon e_N} + 1\right) \qquad \text{(1C-16)}$$

and

$$\Delta P_{IF,Lin}(dBm) = 20 \log\left(\frac{\Delta AGC}{A_\Delta A_\varepsilon e_N} + 1\right) \qquad \text{(1C-17)}$$

The IF output may be divided (power split) to drive two or more circuits. Provided all circuitry is linear, the change in power-split output, $\Delta P_o(dBm)$, will be the same as $\Delta P_{IF}(dBm)$.

---

*If the AGC voltage does not start at zero volt, $\Delta AGC/2$ must be added to the voltage at which the AGC action does start.

Appendix 1D

AGC GAIN CALCULATION

The loop gain of the AGC loop is dependent on the AGC gain of the variable gain IF amplifier. The AGC gain is defined as the change in IF output voltage for a given change in AGC voltage, or (Figure 1-2).

$$A_{AGC} = \frac{\Delta e_{IF}(PP)}{\Delta AGC} \left(P_{in}(dBm) \right) = \text{constant} \qquad (1D\text{-}1)$$

where

$$\Delta e_{IF}(PP) = e_{IF,max}(PP) - e_{IF,min}(PP)$$

$$\Delta AGC = AGC_{max} - AGC_{min}$$

$$AGC_{min} = \text{AGC voltage for minimum gain}$$

$$AGC_{max} = \text{AGC voltage for maximum gain}$$

The IF output power is

$$P_{IF}(dBm) = X(AGC) \qquad (1D\text{-}2)$$

The IF output voltage in terms of output power is (Appendix 1A)

$$e_{IF}(PP) = (0.63) \, 10^{\frac{P_{IF}(dBm)}{20}} \qquad (1D\text{-}3)$$

or, substituting Equation (1D-2),

$$e_{IF}(PP) = (0.63) \, 10^{\frac{X(AGC)}{20}} \qquad (1D\text{-}4)$$

96

The change in amplifier output voltage now may be written as

$$\Delta e_{IF}(PP) = 0.63 \left[10^{\frac{X(AGC)_{max}}{20}} - 10^{\frac{X(AGC)_{min}}{20}} \right] \qquad (1D\text{-}5)$$

or

$$\Delta e_{IF}(PP) = 0.63 \left(10^{\frac{X(AGC)_{min}}{20}} \right) \frac{10^{\frac{X(AGC)_{max}}{20}}}{10^{\frac{X(AGC)_{min}}{20}}} - 1 \qquad (1D\text{-}6)$$

or, using Equation (1D-4) and simplifying,

$$\Delta e_{IF}(PP) = e_{IF,min}(PP) \left(10^{\frac{X\Delta AGC}{20}} - 1 \right) \qquad (1D\text{-}7)$$

The parenthetical term in Equation (1D-7) is of the general form a^Y. Expanding this exponential function,

$$a^Y = 1 + Y \ln(a) + \left[\frac{Y \ln(a)}{2!} \right]^2 + \left[\frac{Y \ln(a)}{3!} \right]^3 + \dots \qquad (1D\text{-}8)$$

Provided Y is small (Y < 0.5), Equation (1D-8) simplifies to

$$a^Y \simeq 1 + Y \ln(a) \qquad (1D\text{-}9)$$

Equation (1D-7) now becomes

$$\Delta e_{IF}(PP) = e_{IF,min}(PP) \left[\frac{X\Delta AGC}{20} (\ln 10) \right] \qquad (1D\text{-}10)$$

or

$$\Delta e_{IF}(PP) = 0.115 \; X\Delta AGC \; e_{IF,min}(PP) \qquad (1D\text{-}11)$$

97

Now Equation (1D-11) may be solved to find the AGC gain,

$$A_{AGC} = 0.115Xe_{IF,min}(PP)$$

(1D-12)

and this equation is valid, to within 10%, for

$$\frac{X\Delta AGC}{20} < 0.5$$

(1D-13)

The $e_{IF,min}(PP)$ term may be replaced by $e_{IF}(PP)$; thus,

$$A_{AGC} = 0.115Xe_{IF}(PP)$$

(1D-14)

or, for $e_{IF}(PP)$ in dBm (Appendix 1A),

$$A_{AGC} = 72.4 \times 10^{-3} \times 10^{\frac{P_{IF}(dBm)}{20}}$$

(1D-15)

Appendix 1E

LOOP RISE TIME CALCULATIONS

The loop rise time for the integrator AGC loop (Figure 1-2b) will now be found. The general method will be to solve for the AGC voltage in terms of input power. This equation can then be differentiated to find the change in AGC voltage with change in input power.

The AGC error voltage, e_ε (Figure 1-2a), is zero under static conditions; however, if the input power is increased, the error voltage will increase and fall back toward zero as the loop nulls. This error voltage may be given as

$$e_\varepsilon = A_v A_\Delta e_D - A_\Delta E_{Ref} \qquad \text{(1E-1)}$$

The value for e_D may be given as $e_{D,N}$, the normalized or static value, plus Δe_D, the change in e_D due to a change in input power. The static value for e_D may be given, for a square law detector, as (Appendix 1B)

$$e_{D,N} = K_{SL} \, 10^{\dfrac{P_{PD,N}(dBm)}{10}} \quad (mV) \qquad \text{(1E-2)}$$

The value for Δe_D is

$$\Delta e_D = [A_{D,SL}(V/dBm)][\Delta P_{PD}(dBm)] \quad (mV) \qquad \text{(1E-3)}$$

or, from Appendix 1B (Figure 1B-2),

$$\Delta e_D = 0.23 \, e_{D,N} \, \Delta P_{PD}(dB) \quad (V) \qquad \text{(1E-4)}$$

where

$e_{D,N}$ = normalized detector output voltage, volts

The instantaneous detector output may now be given as

$$e_D = e_{D,N} + 0.23\, e_{D,N}\, [\Delta P_{PD}(dB)] \qquad \text{(1E-5)}$$

The value for $\Delta P_{PD}(dB)$ may be given as

$$\Delta P_{PD}(dB) = P_{PD}(dBm) - P_{PD,N}(dBm) \qquad \text{(1E-6)}$$

since

$$P_{PD}(dBm) = P_{in}(dBm) + A_{IF}(dB) + A_{PD}(dB) \qquad \text{(1E-7)}$$

and (Figure 1-3)

$$A_{IF}(dB) = A_o(dB) - X(AGC) \qquad \text{(1E-8)}$$

$$P_{PD}(dBm) = P_{in}(dBm) + A_o(dB) - X(AGC) + A_{PD}(dB) \qquad \text{(1E-9)}$$

Equation (1E-5) now becomes

$$e_D = e_{D,N}\,\{1 + 0.23[P_{in}(dBm) + A_o(dB) - X(AGC)$$

$$+ A_{PD}(dB) - P_{PD,N}(dBm)]\} \quad (V) \qquad \text{(1E-10)}$$

The error voltage may now be written as

$$e_\varepsilon = A_v A_\Delta\left(e_{D,N}\left\{1 + 0.23\left|P_{in}(dBm) + A_o(dB) - X(AGC)\right.\right.\right.$$

$$\left.\left.\left. + A_{PD}(dB) - P_{PD,N}(dBm)\right|\right\} - \frac{E_{Ref}}{A_v}\right) \quad (V) \qquad \text{(1E-11)}$$

The integrator output is

$$AGC(S) = \frac{e_\varepsilon(S)A_\varepsilon}{RCS} \qquad (1E\text{-}12)$$

Substituting Equation (1E-12) into (1E-11) and solving for $AGC(S)$,

$$AGC(S) = A_v A_\Delta A_\varepsilon \left(e_{D,N} \left\{ 1 + 0.23 \left| P_{in}(dBm) + A_o(dB) + A_{PD}(dB) \right. \right. \right.$$

$$\left. \left. \left. - AGC(S) - P_{PD,N}(dB) \right| - \frac{E_{Ref}}{A_v} \right\} \right) / (RCS + 0.23\, XA_v A_\Delta A_\varepsilon e_{D,N}) \qquad (1E\text{-}13)$$

To find the change in AGC voltage for a given change in input power, Equation (1E-13) is differentiated with respect to $P_{in}(dBm)$, giving

$$\frac{\Delta AGC(S)}{\Delta P_{in}(dB)(S)} = \frac{0.23\, A_v A_\Delta A_\varepsilon e_{D,N}}{RCS + 0.23\, XA_v A_\Delta A_\varepsilon e_{D,N}} \qquad (1E\text{-}14)$$

or

$$\Delta AGC(S) = \frac{0.23\, A_v A_\Delta A_\varepsilon e_{D,N}}{RCS + 0.23\, XA_v A_\Delta A_\varepsilon e_{D,N}} = \Delta P_{in}(dB)(S) \qquad (1E\text{-}15)$$

Assuming a step change in input power,

$$\Delta AGC(S) = \frac{0.23\, A_v A_\Delta A_\varepsilon e_{D,N}}{RCS + 0.23\, XA_v A_\Delta A_\varepsilon e_{D,N}} \left| \frac{\Delta P_{in}(dB)(S)}{S} \right| \qquad (1E\text{-}16)$$

The inverse Laplace transform of Equation (1E-16) is

$$AGC(t) = \frac{\Delta P_{in}(dB)}{X} \left[1 - \exp - \left(\frac{t/RC}{0.23\, XA_v A_\Delta A_\varepsilon e_{D,N}} \right) \right] \qquad (1E\text{-}17)$$

The time constant for Equation (1E-17) is

$$T = \frac{RC}{0.23\, XA_v A_\Delta A_\varepsilon e_{D,N}} \qquad (1E\text{-}18)$$

The 10 to 90% rise time, τ_r, may be approximated as

$$\tau_r \simeq 2.2T \qquad (1E\text{-}19)$$

Thus

$$\tau_{r,SL} = \frac{9.56\ RC}{XA_v A_\Delta A_\varepsilon e_{D,N}} \qquad (1E\text{-}20)$$

or, since

$$A_v e_{D,N} = e_N \qquad (1E\text{-}21)$$

$$\tau_{r,SL} = \frac{9.56\ RC}{XA_v A_\varepsilon e_N} \qquad (1E\text{-}22)$$

Using the same procedure, the loop rise time for the linear detector is

$$\tau_{r,Lin} = \frac{19.13\ RC}{XA_v A_\varepsilon e_N} \qquad (1E\text{-}23)$$

The loop rise time for the low-pass-filter AGC loop is similar to that for the integrator AGC loop. One main difference, however, is that the error voltage is zero for the integrator AGC loop under static conditions, and is finite for the low-pass-filter loop. The error voltage is small for a well-designed low-pass-filter AGC loop, and may usually be neglected. The AGC voltage, for a square law detector, may be given as

$$\Delta AGC(S) = \frac{0.23\ A_v A_\Delta A_\varepsilon e_{D,N}}{1 + RCS + 0.23\ XA_v A_\Delta A_\varepsilon e_{D,N}}\ \Delta P_{in}(dB)(S) \qquad (1E\text{-}24)$$

The AGC voltage for a step change in input power is

$$\Delta AGC(S) = \frac{0.23\ A_v A_\Delta A_\varepsilon e_{D,N}}{RCS + 0.23\ XA_v A_\Delta A_\varepsilon e_{D,N} + 1}\left[\frac{\Delta P_{in}(dB)}{S}\right] \qquad (1E\text{-}25)$$

102

If

$$0.23\, X A_v A_\Delta A_\varepsilon e_{D,N} \gg 1 \qquad (1E\text{-}26)$$

$$\Delta AGC(S) = \frac{0.23\, A_v A_\Delta A_\varepsilon e_{D,N}}{RCS + 0.23\, X A_v A_\Delta A_\varepsilon e_{D,N}} \left[\frac{\Delta P_{in}(dB)}{S} \right] \qquad (1E\text{-}27)$$

which is exactly the same equation as the integrator AGC loop, Equation (1E-16). Thus, provided the conditions of Equation (1E-20) are met, the low-pass-filter AGC loop has the same loop rise time as the integrator AGC loop, Equations (1E-22) and (1E-23).

Nomenclature

A	conventional gain term
A_{AGC}	dynamic AGC gain
A_D	dynamic detector gain
$A_{D,Lin}$	dynamic detector gain for linear detector
$A_{D,SL}$	dynamic detector gain for square law detector
AGC	automatic gain control
$A_{IF}(\text{dB})$	IF amplifier gain, in dB
A_{Int}	integrator frequency dependent gain
$A_o(\text{dB})$	maximum IF amplifier gain, in dB
A_{PD}	predetector amplifier gain
A_v	video amplifier voltage gain
A_Δ	differencing amplifier gain
A_ε	error amplifier voltage gain
C	capacitance
CR	compression ratio
CR_{IF}	compression ratio of the IF
CR_{Vid}	compression ratio of the video
CW	continuous wave
D	integrator or low-pass-filter loop update duty cycle
dBm	decibel referenced to 1 milliwatt
dBV	decibel referenced to 1 volt
e_d	maximum ac signal across a diode for a given gain error
E_{Ref}	AGC reference voltage
e_D	detector output voltage
$e_{D,Lin}$	output of linear detector

104

$e_{D,SL}$	output of square law detector
$e_{D,N}$	normalized detector output voltage
$e_{IF}(PP)$	peak-to-peak IF amplifier output voltage
e_N	normalized video output voltage
$e_{PD}(PP)$	peak-to-peak detector input voltage
e_{rms}	root mean squared voltage
e_V	video amplifier output voltage
EW	electronic warfare
f	frequency in Hertz
fcs	conical scan frequency
$f_{3dBV}(LPF)$	low-pass-filter 3-dBV frequency response
$f_{3dB,Lin}$	3-dB frequency response for linear detector
$f_{3dB,SL}$	3-dB frequency response for square law detector
IMR	input modulation reduction
IF	intermediate frequency
I_F	diode forward current
I_s	diode reverse saturation current
I_T	constant current source in milliamperes
j	$\sqrt{-1}$
K	Boltzmann's constant
K_1, K_2	logarithmic video amplifier constants
K_{Lin}	linear detector diode constant
K_{SL}	square law detector diode constant
K_{P1}, K_{P2}	PIN diode constants
LG	loop gain
$LG(f)$	frequency dependent loop gain
LG_{Lin}	loop gain for linear detector
LG_{SL}	loop gain for square law detector

LPF	low-pass filter
LS	logarithmic amplifier slope, in V/dB
m	diode slope (ΔV_F/decade of IF)
Me_N	normalized video modulation
M_o	output modulation
M_I	input modulation
MP_{IF}	IF output modulation
MP_{in}	IF input modulation
$P_{IF,max}$(dBm)	maximum IF amplifier signal output power, in dBm, for AGC action
$P_{IF,min}$(dBm)	minimum IF amplifier signal output power, in dBm, for AGC action
P_{in}(dBm)	signal input power, in dBm
$P_{in,max}$(dBm)	maximum signal input power, in dBm, for AGC action
$P_{in,min}$(dBm)	minimum signal input power, in dBm, for AGC action (sometimes referred to as AGC delay)
P_o(dBm)	signal output power, in dBm
P_{oN}(dBm)	normalized signal output power, in dBm
P_{PD}(dBm)	detector input power, in dBm
PRF	pulse repetition frequency
PRI	pulse repetition interval
q	electron charge
R_C	collector resistor
R_E	dc emitter resistance
RF	radio frequency
R_s	diode bulk resistance
r_d	diode dynamic resistance
r_{PIN}	PIN diode dynamic resistance

S Laplacian S

SL square law

T absolute temperature (Kelvins)

T_{pulse} integrator or low-pass-filter time constant

T_u integrator or low-pass-filter AGC loop update time

V_F diode forward voltage

V_T KT/q

X slope of variable gain amplifier, dB/V

$Y_{AGC}(f)$ conical scan filter and sample/hold transfer function

Z_F operational amplifier feedback impedance

ΔAGC change in AGC voltage into IF amplifier

$\Delta AGC'$ change in AGC voltage at error amplifier output

$\Delta e_{IF}(PP)$ peak-to-peak change in IF amplifier output voltage

$\Delta e_N(dB)$ change in normalized video output in dB

$\Delta e_{N,Lin}(dB)$ change in normalized video voltage, in dB, for linear detector

$\Delta e_{N,SL}(dB)$ change in normalized video voltage, in dB, for square law detector

$\Delta e_{PD}(PP)$ peak-to-peak change in detector input voltage

$\Delta e_{\varepsilon}(PP)$ peak-to-peak change in error voltage

$\Delta P_{IF}(dB)$ change in IF amplifier output power, in dB

$\Delta P_{IF,Lin}(dB)$ change in IF amplifier output power, in dB, for linear detector

$\Delta P_{IF,SL}(dB)$ change in IF amplifier output power, in dB, for square law detector

$\Delta P_{in}(dB)$ change in input power, in dB

$\Delta P_o(dB)$ change in output power, in dBm

$\Delta P_{o,Lin}(dB)$ change in output power, in dBm, for linear detector

$\Delta P_{o,SL}(\mathrm{dB})$	change in output power, in dBm, for square law detector
$\Delta P_{PD}(\mathrm{dB})$	compressed output variation
η	diode constant
θ_{PR}	conical scan plane rotation due to AGC loop
τ_r	loop rise time
$\tau_{r,Lin}$	loop 10 to 90% rise time for linear detector
$\tau_{r,SL}$	loop 10 to 90% rise time for square law detector
ϕ	conical scan modulation phase shift, in degrees
ΔV_F	diode forward voltage match necessary for a given gain match

Bibliography

This bibliography is arranged in chronological order. My reason for this is to pay tribute to those whose works formed the foundation on which we build.

Oliver, B. M. "Automatic Volume Control as a Feedback Problem." *Proceedings of the IRE*, April 1948, pp. 466-473.

Field, J. C. G. *The Design of Automatic-Gain-Control Systems for Auto-Tracking Radar Receivers.* The Institution of Electrical Engineers, Monograph No. 258R. October 1957, pp. 93-108.

Victor, W. K. and Brockman, M. H. "The Application of Linear Servo Theory to the Design of AGC Loops." *Proceedings of the IRE*, February 1960, pp. 234-238.

Rheinfelder, W. A. *Designing Automatic Gain Control Systems*; *Part-1 Design Parameters*, EEE, December 1964, pp. 43-47. *Part 2 - Circuit Design* - EEE, January 1965, pp. 53-57.

Tausworthe, R. C. *The Design of Synchronous-Detector AGC Systems.* Jet Propulsion Laboratory, Pasadena, CA, 15 February 1966, pp. 39-43. (Technical Report No. 32-819, publication UNCLASSIFIED.)

Senior, Edwin W. *Automatic Gain Control Theory for Pulsed and Continuous Signals.* Stanford Electronics Laboratories Report SU-SEL-67-054, June 1967.

Hughes, R. S. *Semiconductor Variable Gain and Logarithmic Video Amplifiers.* Continuing Education Institute, Inc., 1967.

109

Ossoff, A. "Design of a Solid State IF AGC System for Pulsed Carrier Microwave Receivers." *The Microwave Journal*, July 1967, pp. 43-48.

Hughes, R. S. "Diodes Make Good Gain-Control Devices." *Electronic Design*, 3 February 1970, pp. 54-57.

Hughes, R. S. "Vary Gain Electronically." *Electronic Design*, 27 May 1971, pp. 78-81.

Ohlson, J. E. "Exact Dynamics of Automatic Gain Control," *IEEE Transactions on Communications*, January 1974, pp. 72-75.

Moskowitz, J. "Linear Feedback AGC with Input-Level-Invariant Response Times." *Journal of the Acoustical Society of America*, Vol. 62, No. 6, December 1977, pp. 1449-1456.

Hughes, R. S. "Design Automatic Gain Control Loops the Easy Way." *EDN*, 5, October 1978, pp. 123-128.

Porter, J. "AGC Loop Design Using Control System Theory," *RF Design*, June 1980, pp. 27-32.

Xu, H. and Smith, S. H. "AGC Handles Transients of Pulsed Radar Signal," *Microwaves and RF*, May 1986, pp. 183-190.

Naval Weapons Center. *Effect of Automatic Gain Control and Sample/Hold Characteristics on Plane Rotation of Certain Conical Scan Radar Receivers*, by Richard Smith Hughes. China Lake, CA, NWC, August 1983. (NWC TP 6454, publication UNCLASSIFIED.)

Chapter 2

AUTOMATIC NOISE TRACKING LOOPS

Virtually all radar and electronic warfare systems employ a signal threshold to start the necessary system timing process (signal sample/hold, analog-digital conversion, etc.). The signal threshold voltage must be large enough to prevent thresholding on noise, but not so large as to prevent excessive loss in signal sensitivity. This chapter will discuss signal thresholds from a basic, practical standpoint. The bibliography at the end of this chapter lists many excellent references on signal thresholding. It is not the intent of this chapter to be an all-encompassing thresholding tutorial, but rather a discussion of the need for, and basic design of, an automatic signal threshold that adjusts its voltage to keep the false alarm rate (signal crossings due to noise) constant, despite excess noise generated in the receiver (due to changes in gain due to automatic gain control (AGC)).

Figure 2-1 illustrates the basic configuration that will be presented. The received signal is detected and amplified by the video amplifier. The resultant signal and noise drive the signal threshold comparator.

The signal threshold $(V_T|_S)$ is generally set for a given false alarm rate (FAR) in the absence of signal. The false alarm rate is dependent on individual system requirements; however, values generally range from one noise crossing every ten seconds (FAR = 0.1) to ten noise crossings every second (FAR = 10).

111

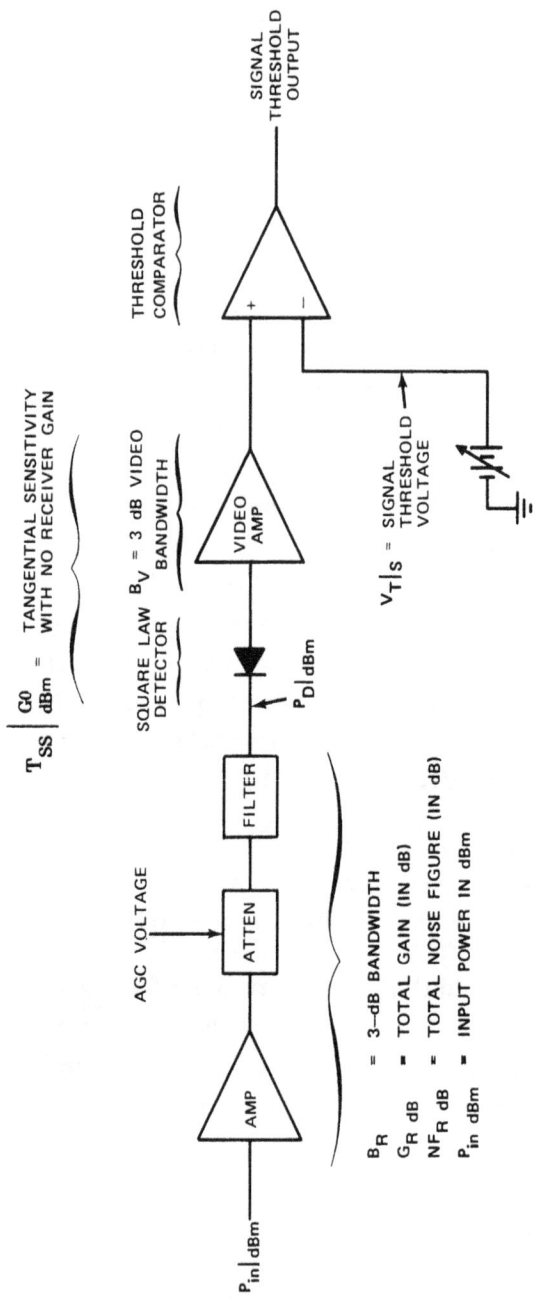

FIGURE 2-1. Basic Receiver Configuration.

$P_{in}|dBm$

AMP

AGC VOLTAGE

ATTEN

FILTER

$T_{SS}\left|\begin{array}{c}G0\\dBm\end{array}\right.$ = TANGENTIAL SENSITIVITY WITH NO RECEIVER GAIN

SQUARE LAW DETECTOR

B_V = 3 dB VIDEO BANDWIDTH

$P_D|dBm$

VIDEO AMP

$V_T|s$ = SIGNAL THRESHOLD VOLTAGE

THRESHOLD COMPARATOR

+

−

SIGNAL THRESHOLD OUTPUT

B_R = 3–dB BANDWIDTH
G_R dB = TOTAL GAIN (IN dB)
NF_R dB = TOTAL NOISE FIGURE (IN dB)
P_{in} dBm = INPUT POWER IN dBm

112

The signal's threshold voltage determines the receiver's sensitivity.* Thus, a large threshold voltage (for a low false alarm rate) requires a large receiver input for signal detection (thus a loss in sensitivity). Signal sensitivity is defined here as that input signal required to give an 80% count of the input PRF (i.e., for a PRF of 2 kHz, the 80% probability of detection $(P_d \mid_{dBm}^{80\%})$ occurs when the pulse count is 1.6 kHz).

The video amplifier output noise (for unity receiver gain) is a function of the detector, video amplifier noise figure, video bandwidth (B_V), and the video amplifier gain (A_V). The sensitivity obtained from such a configuration (commonly called a crystal video receiver) can be greatly increased by increasing the receiver gain (Appendix 2A). The video noise is independent of receiver gain until the receiver noise into the detector becomes dominant. Further increases in gain continue to provide an increase in sensitivity; however, the noise at the output of the video amplifier also increases due to increases in IF gain. Thus, the signal threshold must be increased to maintain the required false alarm rate. Setting the signal threshold (for the required false alarm rate) at large receiver gains results in a loss in sensitivity and dynamic range as the receiver gain, thus, video output noise, decreases.

Consider a receiver that has a bandwidth (B_R) of 120 MHz, a noise figure of 5 dB, and a video bandwidth (B_V) of 7.7 MHz (Appendix 2A). Figure 2-2 shows the sensitivity ($P_d \mid_{dBm}$) and signal threshold voltage ($V_T \mid_S$) for the signal threshold set for a false alarm rate equal to one for several values of G_R. The sensitivity with no receiver gain is −38 dBm (with V_T set for a false alarm rate of one). The sensitivity increases 1 dB for each dB increase in receiver gain, as expected. The video amplifier output noise is constant for receiver gains less than about 30 dB; however, as the gain is increased beyond this

---

*The bibliography at the end of this chapter contains an extensive listing of sensitivity-related material.

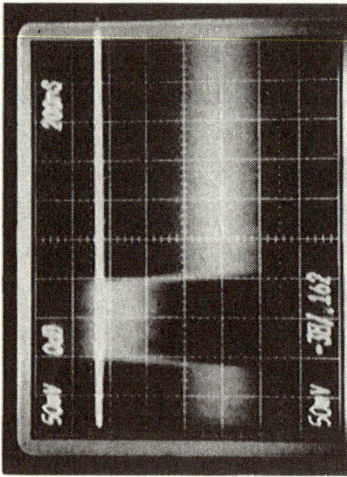

(a) $G_R = 0$ (Crystal Video Receiver),
$V_T = 0.162$ V, $Pd|_{dBm}^{80\%} = -38$,
rms noise = 33 mV

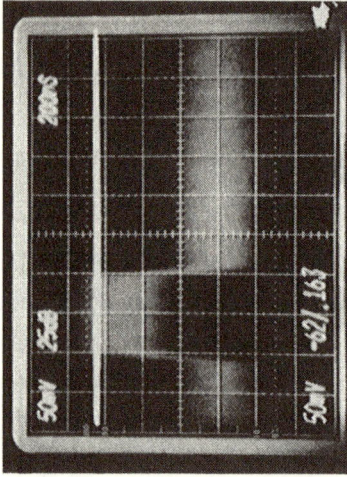

(b) $G_R = 25$ dB, $V_T = 0.163$ V,
$Pd|_{dBm}^{80\%} = -62$,
rms noise = 33 mV

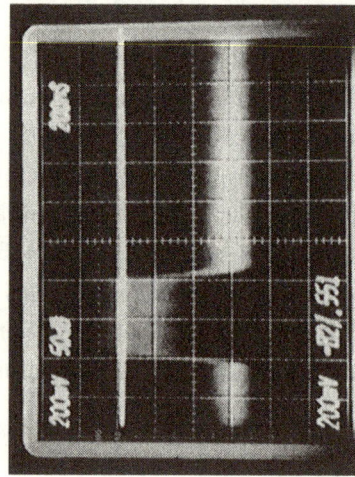

(c) $G_R = 50$ dB, $V_T = 0.551$ V, $Pd|_{dBm}^{80\%} = -82$,
rms noise = 73 mV

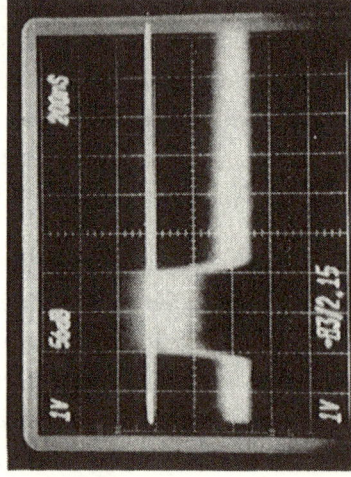

(d) $G_R = 56$ dB, $V_T = 2.15$ V, $Pd|_{dBm}^{80\%} = -83$,
rms noise = 277 mV

FIGURE 2-2. Effect of Receiver Gain on Sensitivity and Signal Threshold (V_T)

114

value, the video noise increases, necessitating an increase in V_T to keep a unity false alarm rate (Figures 2-2c and 2-2d). The loss in dynamic range may be given, for a square law detector (Appendix 1B), as (Appendix 2A, Equation (2A-19))

$$DR\Big|^{Loss}_{dBm} = 10 \, Log \, \frac{V_T\Big|^A_S}{V_T\Big|^{G0}_S} \tag{2-1}$$

where

$DR\Big|^{Loss}_{dBm}$ = loss in input dynamic range

$V_T\Big|^A_S$ = signal threshold needed for required false alarm rate at a particular receiver gain

$V_T\Big|^{G0}_S$ = signal threshold needed for required false alarm rate for no receiver gain

Thus, setting $V_T|_S$ at the maximum expected receiver gain results in a significant loss in input dynamic range as the receiver gain is decreased.

Figure 2-3 shows the signal sensitivity for two fixed thresholds: (1) $V_T|_S$ = 2.15, the value necessary for a false alarm rate of one for a receiver gain of 56 dB, and (2) $V_T|_S$ = 0.551, the value necessary for a false alarm rate of one for a receiver gain of 50 dB. The loss in input dynamic range due to $V_T|_S$ will now be found.

A. Signal threshold voltage ($V_T|_S$) set for a false alarm rate of one at a receiver gain of 56 dB is 2.15 volts (Figure 2-2d). The measured sensitivity (for $G_R = 0$) for this threshold setting is -26 dBm (Figure 2-3a). The optimum sensitivity for $V_T|_S$ = 0.162 is -38 dBm (Figure 2-2a). Thus, there is a loss in input dynamic range (and sensitivity) of

$$DR\Big|^{Loss}_{dBm} = -26 - (-38) = 12 \, dB \tag{2-2}$$

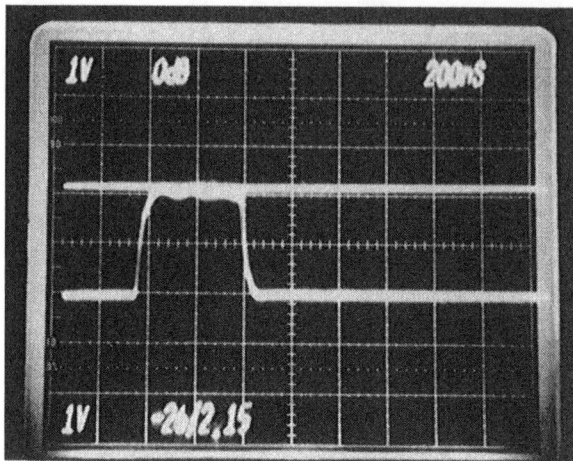

(a) $G_R = 0$, $V_T|_S = 2.15$ V (see Figure 2-2d),
$P_d\big|^{80\%}_{dBm} = -26$.

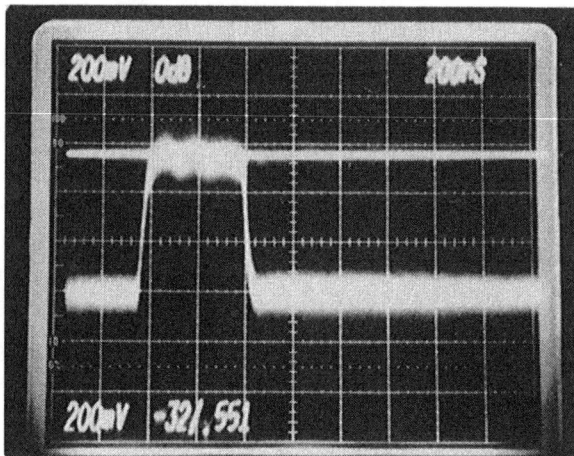

(b) $G_R = 0$, $V_T|_S = 0.551$ V (see Figure 2-2c),
$P_d\big|^{80\%}_{dBm} = -32$.

FIGURE 2-3. Effect of $V_T|_S$ on Sensitivity.
$B_R = 120$ MHz, $B_V = 7.7$ MHz, $NF_R = 5$ dB,
PW = 0.5 μsec.

The predicted loss in input dynamic range is (Equation (2-1))

$$DR \bigg|_{dBm}^{Loss} = 10 \text{ Log } \frac{2.51}{0.162} = 11.3 \qquad (2\text{-}3)$$

B. Signal threshold voltage set for a false alarm rate of one at a receiver gain of 50 dB is 0.551 volt (Figure 2-2c). The measured sensitivity (for $G_R = 0$ dB) for this threshold setting is -32 dBm (Figure 2-3b). The optimum sensitivity for $V_T|_S = 0.162$ is -38 dBm (Figure 2-2a). Thus the loss in input dynamic range is

$$(2\text{-}4)$$

$$DR \bigg|_{dBm}^{Loss} = -32 - (-38) = 6 \text{ dB}$$

The predicted loss is (Equation (2-1))

$$DR \bigg|_{dBm}^{Loss} = 10 \text{ Log } \frac{0.551}{0.162} = 5.3 \text{ dB} \qquad (2\text{-}5)$$

which also is in excellent agreement with the 6-dB measured value.

Figure 2-4 shows the increase in sensitivity and loss in input dynamic range as a function of receiver gain.

Figure 2-5 shows the increase in false alarm rate with increased receiver gain for a signal threshold of 0.162 volt (Figure 2-2a). It is obvious that for receiver gains in excess of about 40 dB, the false alarm rate increase is intolerable. The system engineer must make a decision about decreasing receiver gain and losing sensitivity, or increasing the signal threshold and losing dynamic range. What to do?

If a radar or EW system requires the maximum sensitivity available, and the loss in input dynamic range accompanying a fixed signal threshold cannot be tolerated, a noise riding threshold may be the only solution.

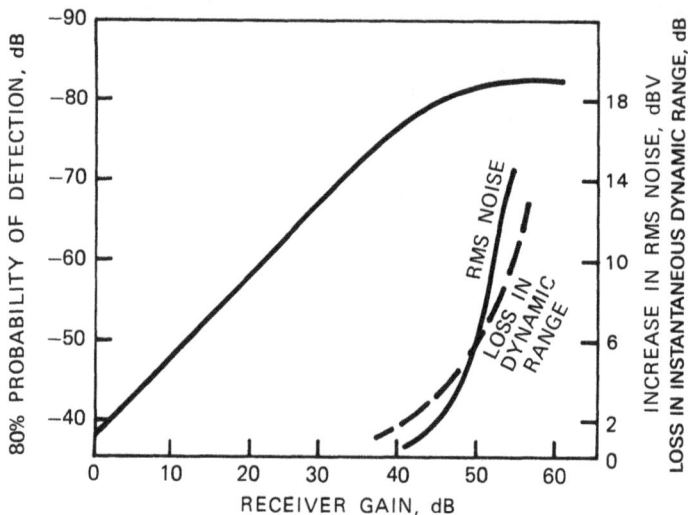

FIGURE 2-4. Effect of Receiver Gain on Sensitivity and Dynamic Range. B_R = 120 MHz, B_V = 7.7 MHz, NF_R = 5 dB, PW = 0.5 μsec, rms noise (gain < 30 dB) = 35 mV, FAR = 1.

One method to obtain a noise riding threshold is illustrated in Figure 2-6. Circuit operation is straightforward: the noise comparator is triggered for noise voltages greater than the noise threshold voltage ($V_T|_N$). The noise comparator output triggers the noise one-shot (to ensure reliable triggering of the noise counter). The resultant noise is counted (provided no signal is present) during the counter enable time (T_C) and digitally converted to an analog voltage (DAC). The DAC analog voltage (which is controlled by the noise count during the noise counter enable time) is compared with a reference voltage ($-V_{Ref}$) and the resulting error voltage (V_ε) is

$$V_\varepsilon = \frac{V_{DAC} - V_{Ref}}{2} \tag{2-6}$$

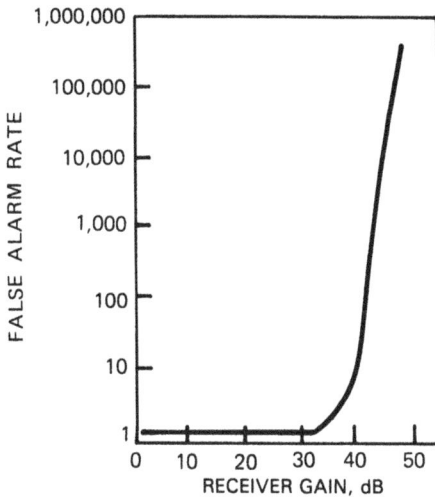

FIGURE 2-5. Increase in False Alarm Rate with Receiver Gain for $V_T = 0.162$, $B_R = 200\,MHz$, $B_V = 7.7\,MHz$, $NF_R = 5\,dB$, $PW = 0.5\,\mu sec.$

which is integrated during the noise integrator enable time (T_I). The noise threshold loop, via feedback, ensures that the noise integrator input voltage (V_ε) is driven to zero. Since there is only one DAC input (noise count) that will ensure $V_{DAC} = V_{Ref}$, the noise comparator output count is constantly independent of input noise amplitude. The noise threshold voltage ($V_T|_N$) is increased by the signal threshold amplifier ($A_V|_{ST}$). The larger the signal threshold amplifier gain, the larger the signal threshold voltage; thus, the lower the signal comparator false alarm rate. To ensure that the noise count is insensitive to the input signal, the noise counter is disabled via the signal threshold comparator. The signal comparator one-shot ensures that the first signal threshold crossing determines the sensitivity. Without this one-shot (or similar digital element), a threshold count well in excess of the input PRF will be obtained for large receiver gains (thus large noise) as illustrated in Figure 2-7.

119

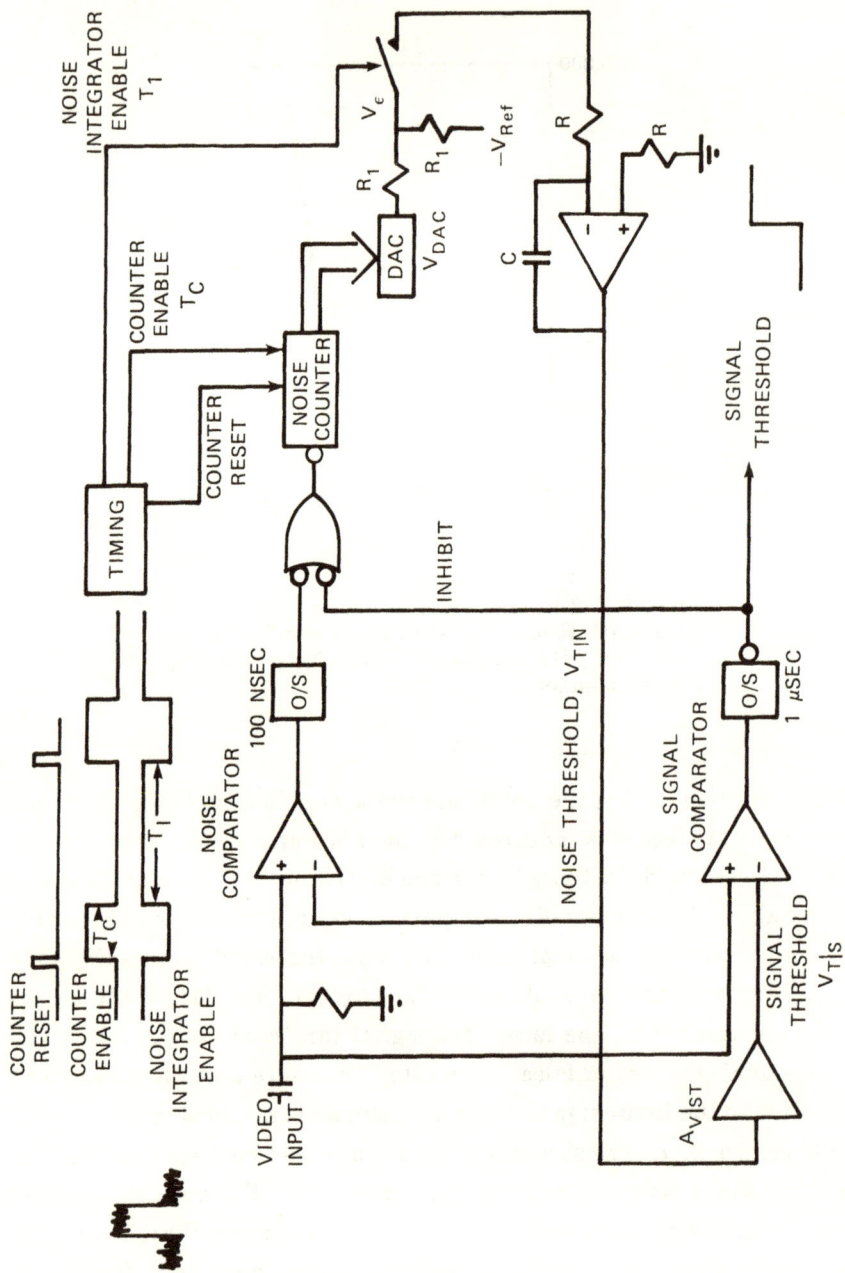

FIGURE 2-6. Noise Riding Threshold.

120

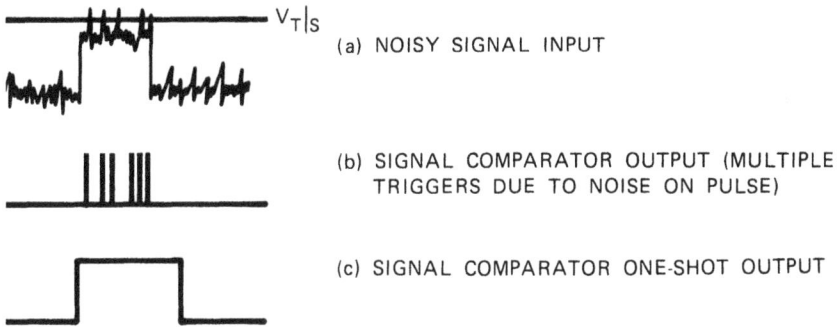

(a) NOISY SIGNAL INPUT

(b) SIGNAL COMPARATOR OUTPUT (MULTIPLE TRIGGERS DUE TO NOISE ON PULSE)

(c) SIGNAL COMPARATOR ONE-SHOT OUTPUT

FIGURE 2-7. Excessive Sensitivity Count and Realistic Count for Noisy Signals.

Appendix 2A shows that for a noise count (false alarm rate) of 250 kHz, the threshold voltage is reasonably insensitive to receiver gain, and this value will be used for the normalized noise threshold count in Figure 2-6. The noise count is only enabled for time, T_C (Figure 2-6), since a count of 250 kHz would require an excessively large DAC. Equation (2-7) presents the relationship between the desired normalized noise threshold count (FAR_N), counter enable time (T_C), and normalized noise count (N)

$$N = [FAR_N \text{ (in kHz)}] \, [T_C \text{ (in } \mu sec)] \times 10^{-3} \qquad (2\text{-}7)$$

or

$$T_C \, (\mu sec) = \frac{N \times 10^3}{FAR_N \text{ (in kHz)}} \qquad (2\text{-}8)$$

where

> N = normalized noise counter count for a
>
> T_C = DAC enable time (in µsec) for a
>
> FAR_N = normalized noise threshold count (in kHz).

Assume the following:

$$FAR_N = 250\,kHz, \quad N = 8, \quad T_C\,(\mu sec) = \frac{8 \times 10^3}{250\,(kHz)} = 32\,\mu sec \tag{2-9}$$

The noise integrator update time (T_I) should be as small as practical to ensure that the noise riding threshold can track noise changes due to changes in receiver gain. A value of $T_I = T_C$ will be used for the example to follow.

The noise riding threshold illustrated in Figure 2-6 will be designed for receiver/detector video amplifier characteristics illustrated in Figures 2-2, 2-3, 2-4, and 2-5.

Design characteristics:

$$FAR_N = 250\,kHz \quad N = 8 \quad T_C = 32\,\mu sec \quad T_I = T_C = 32\,\mu sec$$

Figure 2-8 illustrates the functional noise riding threshold schematic. The three least significant bits for the DAC are grounded and only the four most significant bits are used. The DAC08 has a current output and the voltage integrator error (Figure 2-6) may be represented as shown in Figure 2-9. The noise counter and DAC (Figure 2-7) are configured such that a noise count of eight in 32 µsec (see Equation (2-8)) is obtained. Table 2-1 gives V_ε as a function of noise counts, N. Thus, as the noise count is increased $(N > 8)$, the effective noise integrator error voltage (V_ε) becomes negative, increasing the noise threshold voltage, and thus decreasing N and driving V_ε to zero. The signal threshold amplifier $(A_V|_{ST})$ increases the noise threshold voltage and adjusts for the desired maximum false alarm rate.

$$V_T\big|_S = \left(A_V\big|_{ST}\right)\left(V_T\big|_N\right) \tag{2-10}$$

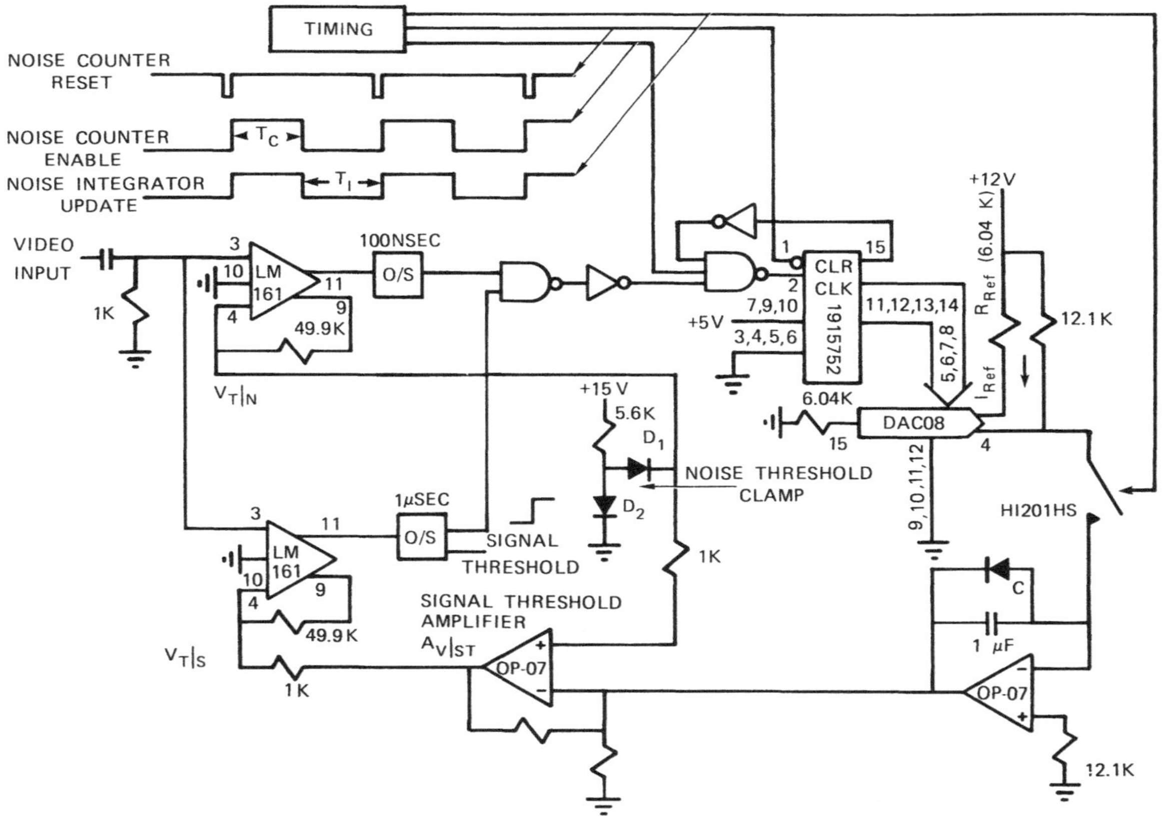

FIGURE 2-8. Functional Schematic: Noise Riding Threshold.

123

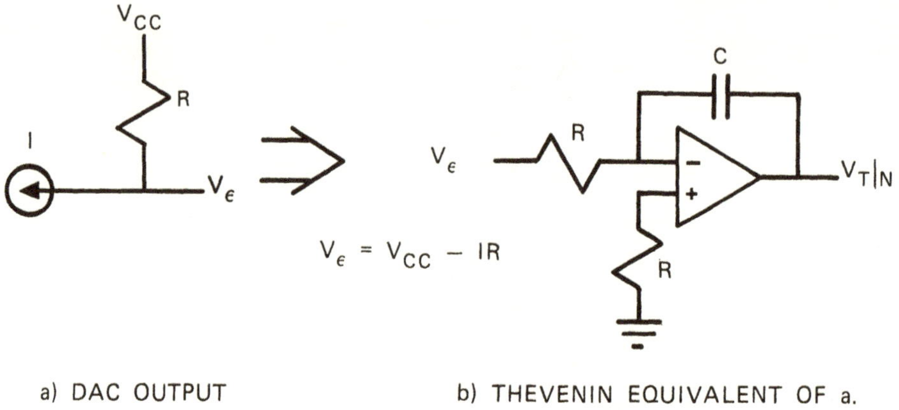

Vcc
R
I
Vε
Vε
R
C
VT|N
Vε = Vcc − IR
R
a) DAC OUTPUT
b) THEVENIN EQUIVALENT OF a.

FIGURE 2-9. DAC Output Voltage.

TABLE 2-1. Noise Error Voltage, V_ε, as a Function of N.

N	DAC input	V_ε
0	0 0 0 0	V_{CC}
8	1 0 0 0	0
15	1 1 1 1	$\approx V_{CC} - I_{Ref}R$ $\approx -V_{CC}$
$I_{Ref} = V_{CC}/R_{Ref}$		

Diodes D_1 and D_2 comprise a lower noise threshold clamp to prevent the noise riding threshold from attempting to lock on negative noise crossings.

Figures 2-10, 2-11, and 2-12 illustrate the measured results. The signal threshold amplifier determines the maximum signal false alarm rate and the results of three values of $A_V|_{ST}$ gain are shown.

A. Figure 2-10 gives the sensitivity $(P_d \big|_{dBm}^{80\%})$ as a function of receiver gain. As can be seen, the data compare quite well with the manually adjusted signal threshold data of Figure 2-4.

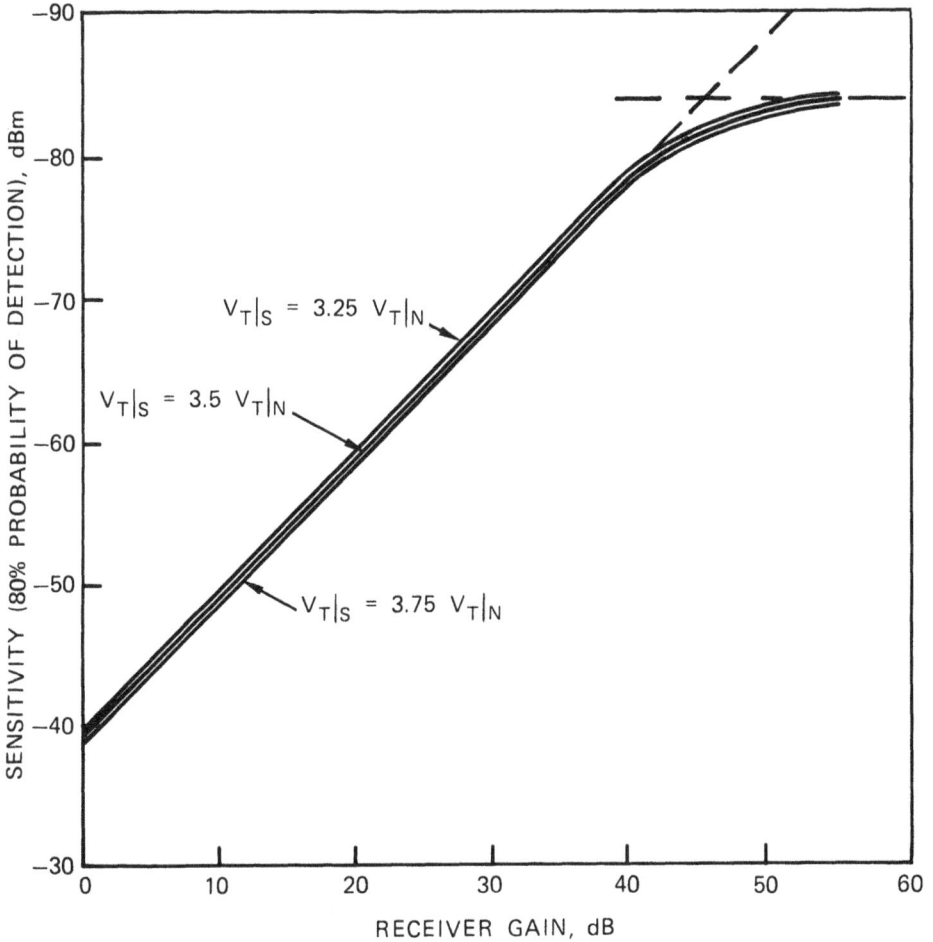

FIGURE 2-10. Sensitivity as a Function of Receiver Gain and Signal Threshold Amplifier Gain.

B. Figure 2-11a shows the noise count as a function of receiver gain, and it is quite insensitive to changes in receiver gain (input noise). Figure 2-11b shows the signal false alarm rate for three values of $A_V|_{ST}$. Increasing $A_V|_{ST}$ decreases the FAR at the expense of sensitivity (Figure 2-10).

(a) Signal False Alarm Rate as a Function of Receiver Gain.

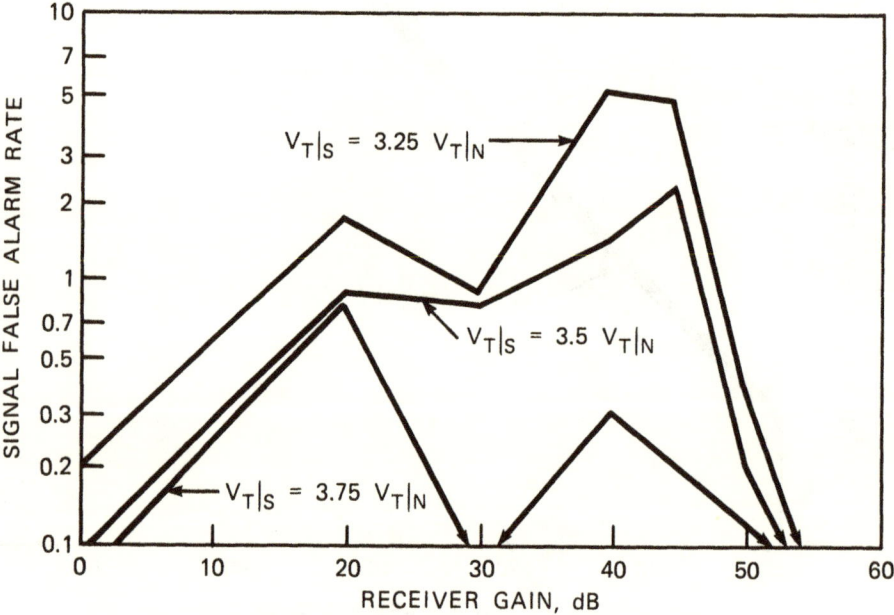

(b) Noise Threshold Count as a Function of Receiver Gain.

FIGURE 2-11. Signal False Alarm Rate and Noise Threshold Count as Functions of Receiver Gain.

C. Figure 2-12 shows the signal threshold and noise threshold voltages as a function of receiver gain.

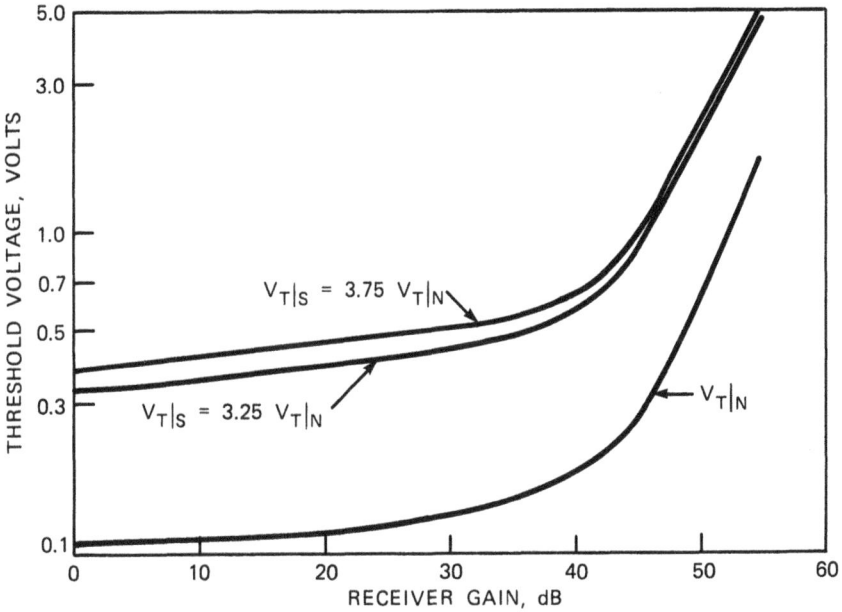

FIGURE 2-12. Threshold Voltages as a Function of Receiver Gain.

A basic model to determine loop closure time for a change in video noise (due to changes in receiver gain) is presented in Figure 2-13. The false alarm rate is modeled as a voltage controlled oscillator having an output frequency that is the normalized false alarm rate (250 kHz for our design example). The oscillator has a slope of

$$S_F = \frac{FAR}{Volt}$$

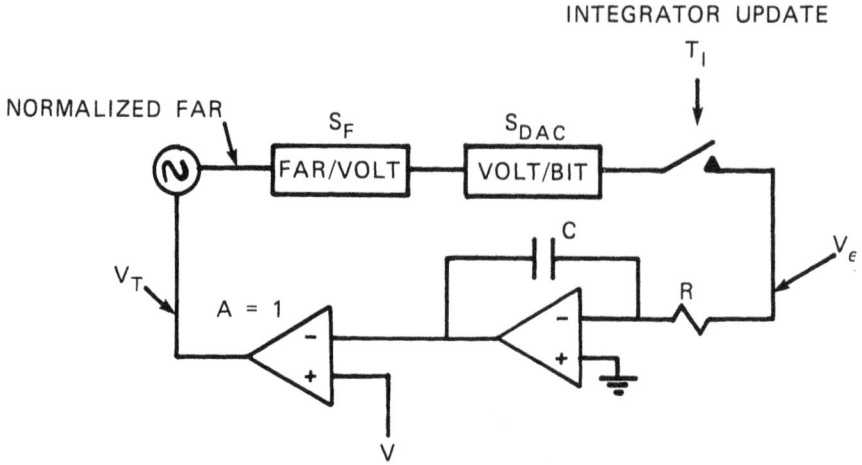

FIGURE 2-13. Basic Noise Riding Threshold Feedback Model.

and is a nonlinear function of voltage (see Figures 2A-9, 2A-11, 2A-13, and 2A-16). Figure 2-14 illustrates false alarm rate as a function of threshold voltage (see Figure 2A-9). At large false alarm rate values (greater than 1,000) the curves can be approximated as

$$FAR = a \exp(-n V_T) \qquad (2\text{-}11)$$

where

$$a = FAR \text{ (for } V_T = 0) \qquad (2\text{-}12)$$

$$n = \left(\frac{1}{V_T} \right) Ln \left(\frac{a}{FAR} \right) \qquad (2\text{-}13)$$

128

and

$$S_F = \frac{d\,FAR}{d\,V_T} = -\,an\,\exp(-n\,V_T)\left(\frac{FAR}{Volt}\right)$$ (2-14)

$$V_T = \left(\frac{1}{n}\right) Ln\left(\frac{a}{FAR}\right)$$ (2-15)

Table 2-2 summarizes the results of Equations (2-11) to (2-15) for the curves of Figure 2-14.

TABLE 2-2. FAR Results from Figure 2-14.

Curve	a (Equation (2-12))	n (Equation (2-13))	FAR (Equation (2-11))	V_T (FAR = 250 kHz) (Equation (2-15))	S_F (FAR = 250 kHz) Equation (2-14)
A	100×10^6	46	100×10^6 $\exp - 46\,V_T$	0.13	-11×10^6 Hz/volt
B	200×10^6	17	200×10^6 $\exp - 17\,V_T$	0.39	-4×10^6 Hz/volt
C	100×10^6	8.2	100×10^6 $\exp - 8.2\,V_T$	0.73	-31×10^3 Hz/volt

The loop closure time will now be found for Figure 2-13.

The loop gain (LG) is

$$LG(f) = \frac{(S_F)(S_{DAC})\,D}{2\pi f\,RC}$$ (2-16)

where

D = integrator update duty cycle (see Equation (1-94))

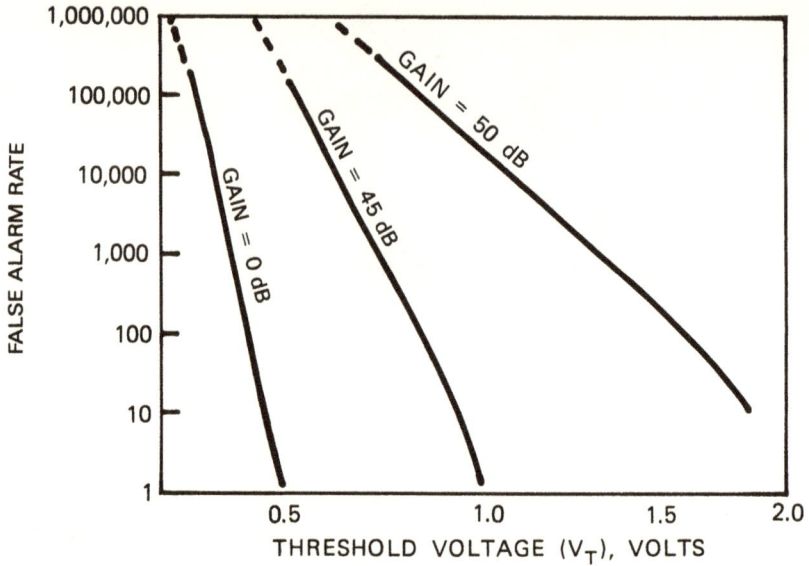

FIGURE 2-14 (see Figure 2A-9). FAR Versus Threshold
Voltage. $B_R = 120$ MHz, $B_V = 7.7$ MHz, $NF_R = 6$.

The loop transfer function is

$$\frac{V_T(f)}{V(f)} = \frac{LG(S)}{1 + LG(S)} = H(S) \qquad (2\text{-}17)$$

substituting Equation (2-16) into (2-17) and simplifying

$$H(S) = \frac{1}{1 + j \dfrac{f S_F S_{DAC} D}{2 \pi RC}} \qquad (2\text{-}18)$$

The 3-dB bandwidth may be given as

$$f_{3dB} = \frac{S_F S_{DAC} D}{2 \pi RC} \qquad (2\text{-}19)$$

130

and the loop rise time may be given as

$$t_r = \frac{0.35}{f_{3dB}} \qquad (2\text{-}20)$$

or, upon substituting Equation (2-19) into (2-20),

$$t_r = \frac{2.2\,RC}{S_F\,S_{DAC}\,D} \qquad (2\text{-}21)$$

and then using the parameters of the design example (Figure 2-7),

$$R = 12.1\,k\Omega \quad C = 1\mu F \quad D \simeq 0.5$$

The DAC scale factor, S_{DAC}, may be approximated as (Table 2-1)

$$S_{DAC} \simeq \frac{2V_{CC}}{N(max)} \qquad (2\text{-}22)$$

or

$$S_{DAC} \simeq \frac{24}{15} = 1.6 \text{ Volts/bit} \qquad (2\text{-}23)$$

and S_F is obtained from Table 2-2.

The loop routine (for a receiver gain of 45 dB) is

$$t_r = \frac{2.2\,(12.1 \times 10^3)(1 \times 10^{-6})}{(4 \times 10^6)(1.6)(0.5)} = 8.3 \times 10^{-9} \qquad (2\text{-}24)$$

To say the least, this predicted loop rise time is ridiculously fast. A much more realistic approach is to assume that the DAC is hard limited for any increase or decrease in false alarm rate ($V_\varepsilon \simeq V_{CC}$ for FAR < 250 kHz and $V_\varepsilon \simeq -V_{CC}$ for FAR > 250 kHz). This is a most reasonable assumption as the noise riding threshold of Figure 2-7 is only counting 15 noise crossings out of 250,000 (in fact, the statistical nature of the noise ensures that V_ε (Figure 2-9) will toggle between $\pm V_{CC}$, with an average value of zero, for a constant input noise and the loop nulled). Since the DAC is limited, the integrator has a constant slew rate for increasing or decreasing input noise. Thus, the change in integrator output voltage may be given as

$$\frac{\Delta V}{\Delta t} = \frac{V_{CC} D}{RC} \tag{2-25}$$

or for the circuit of Figure 2-7

$$\frac{\Delta V}{\Delta t} = \frac{12 (0.5)}{(12.1 \times 10^3)(1 \times 10^{-6})} = 495 \text{ Volt/sec} \tag{2-26}$$

or

$$\frac{\Delta V}{\Delta t} = 0.495 \text{ Volt/sec} \tag{2-27}$$

Referring to Figure 2-14, the threshold voltage difference (for a FAR = 250 kHz) is 0.6 volt as the gain is changed from 50 dB (Figure 2-2(c)) to 0 dB (Figure 2-2(a)). Thus the expected loop closure time is (Equation (2-27))

$$\Delta t = \frac{\Delta V}{0.495} \tag{2-28}$$

or

$$\Delta t = \frac{0.6}{0.495} = 1.2 \text{ msec} \tag{2-29}$$

which is quite close to the measured value shown in Figure 2-15.

The noise riding threshold technique presented is intended only as a starting point. There are many concepts that will achieve automatic signal thresholding and they are limited only by the originality of the designer.

Achieving the optimum sensitivity that can be obtained for a given receiver configuration may necessitate a noise riding threshold if excessive loss in dynamic range is to be avoided. The system engineer must take a close look at system requirements to justify the need for the system level complications of a noise riding threshold.

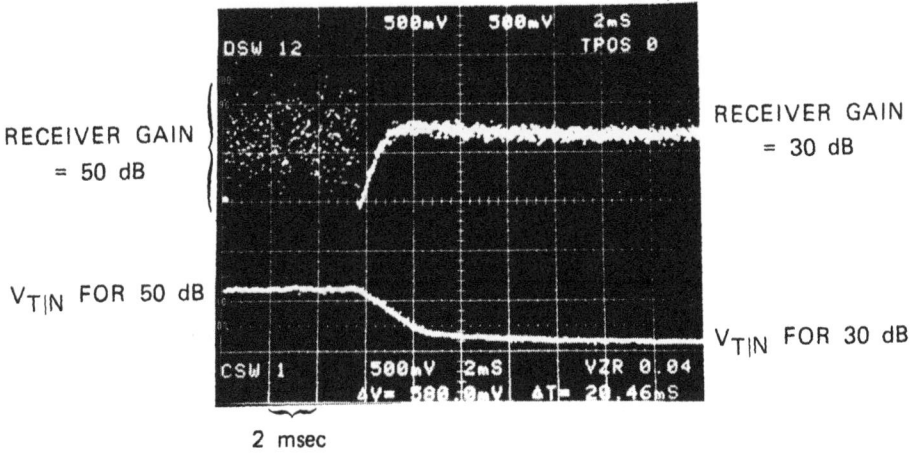

FIGURE 2-15. Noise Riding Threshold Closure Time.
Top trace: video noise (0.5 V/div); bottom trace: noise
threshold (0.5 V/div); horizontal time, 2 msec/div.

References

1. Naval Weapons Center. *Determining Maximum Sensitivity and Optimum Maximum Gain for Detector-Video Amplifiers with RF Preamplification*, by R. S. Hughes. China Lake, CA, NWC, March 1985 (NWC TM 5357, publication UNCLASSIFIED.)

2. Hughes, R. S. "Determining Maximum Sensitivity and Optimum Maximum Gain for Detector-Video Amplifiers with RF Preamplification," *Microwave Journal*, November 1985, pp. 165-178.

3. Tsui, J. "Tangential Sensitivity of EW Receivers," *Microwave Journal*, October 1981, pp. 99-102.

4. Lucas, W. J. "Tangential Sensitivity of a Detector Video System with RF Preamplification," *Proceedings of the IEEE*, Vol. 113, No. 8 (August 1966), pp. 1321-30.

5. Stanford University. *Characteristics of Crystal-Video Receivers Employing RF Preamplification*, by W. E. Ayer. Palo Alto, CA, Stanford Electronic Laboratories, 20 September 1956. (Tech Report No. 150-3, publication UNCLASSIFIED.)

Appendix 2A

THE EFFECT OF RECEIVER GAIN
ON SIGNAL SENSITIVITY AND NOISE

Introducing receiver gain prior to signal detection increases signal sensitivity; however, the video noise increases as the receiver gain is increased beyond a certain point. This increase in noise may well necessitate the need for a noise riding threshold. This appendix presents a basic discussion of signal sensitivity and increased video noise as a function of receiver gain, receiver bandwidth, and video bandwidth, and is based on [1] and [2].

Background

Figure 2A-1 illustrates the basic configuration that will be discussed. The main restrictions that apply for the theory to be presented are (1) the detector is operated in the square law region (Appendix 1B), and (2) the receiver bandwidth (B_R) is at least twice the video bandwidth (B_V) (see Appendix 2B for $B_R \simeq B_V$).

The RF/IF variable gain amplifier amplifies and limits the bandwidth (B_R) of the incoming signal. The resulting signal (and noise) drives the square law detector and video amplifier (see Figure 2A-2 for a brief summary of square law detectors). The resulting signal and noise drives a comparator. If the detected video signal is greater than the signal threshold voltage ($V_T|_S$), the comparator is triggered. A practical method of determining the relationships of the factors that determine signal threshold triggering will be presented.

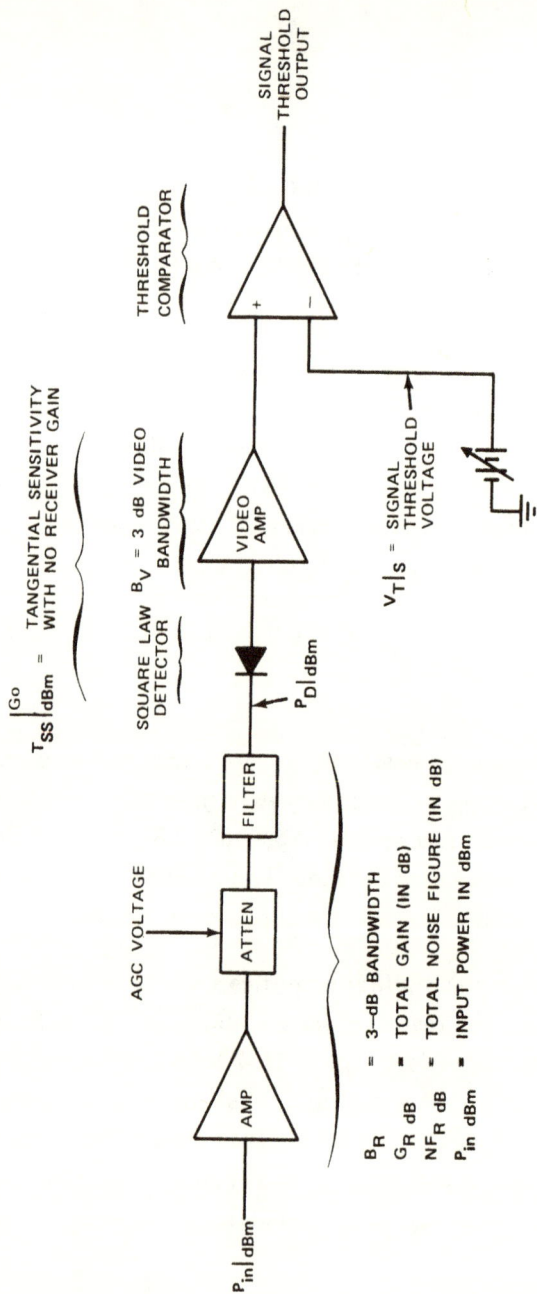

$T_{SS}\big|_{dBm}^{Go}$ = TANGENTIAL SENSITIVITY WITH NO RECEIVER GAIN

SQUARE LAW DETECTOR B_V = 3 dB VIDEO BANDWIDTH

VIDEO AMP

$P_D\big|dBm$

$V_T\big|_S$ = SIGNAL THRESHOLD VOLTAGE

THRESHOLD COMPARATOR

SIGNAL THRESHOLD OUTPUT

AGC VOLTAGE

ATTEN

FILTER

AMP

$P_{in}\big|dBm$

B_R = 3-dB BANDWIDTH
G_R dB = TOTAL GAIN (IN dB)
NF_R dB = TOTAL NOISE FIGURE (IN dB)
P_{in} dBm = INPUT POWER IN dBm

FIGURE 2A-1. Basic Configuration, Detector-Video Receiver.

136

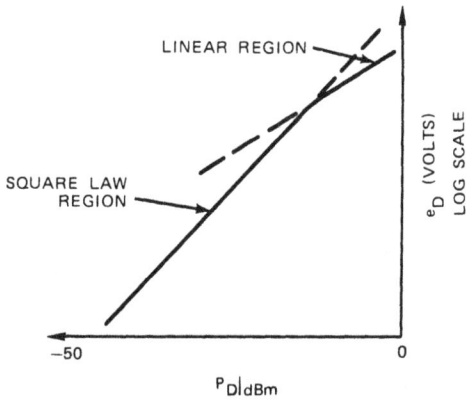

$$\Delta e_D \Big|_{dBv}^{SQ~LAW} = 2\Delta P_D \Big|_{dBm}$$

$$K_{SL} \equiv \frac{e_D~(IN~mV)}{P_D~(IN~mW)} = e_D~(mV)~10^{\frac{-P_D|_{dBm}}{10}} \qquad \left(\frac{mV}{mW}\right)$$

$R_V \equiv$ DETECTOR VIDEO RESISTANCE

$$M \equiv \frac{K_{SL}}{\sqrt{R_V}}$$

FIGURE 2A-2. Square Law Detector Summary.

137

Maximum Tangential Sensitivity and Optimum Maximum RF Gain

If the RF/IF receiver is removed, the input signal ($P_{in}|_{dBm}$) drives the detector directly. An excellent "goodness" factor for the sensitivity of this configuration is the tangential sensitivity, T_{SS}. Tangential sensitivity represents a signal-to-noise power ratio of 6.31 (8 dB) at the video amplifier output, and refers to the condition in which the signal plus noise waveform (on an oscilloscope) protrudes above the noise baseline such that the bottom edge of the combined terms just touches the top edge of the baseline noise as shown in Figure 2A-3. Figure 2A-4 summarizes the measurement technique for determining T_{SS}. [3]

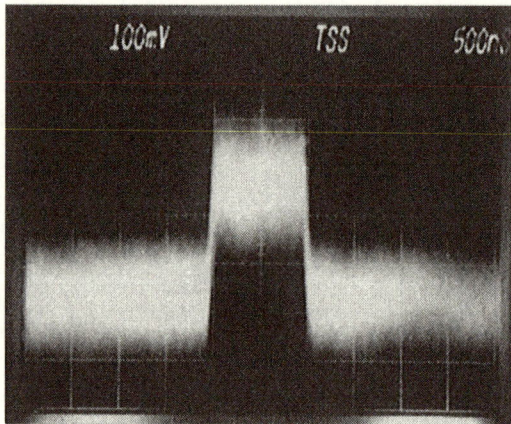

FIGURE 2A-3. Tangential Signal.

RECEIVER SQUARE LAW DETECTOR VIDEO AMPLIFIER

P_{in}

B (INVERTED CHANNEL)

A

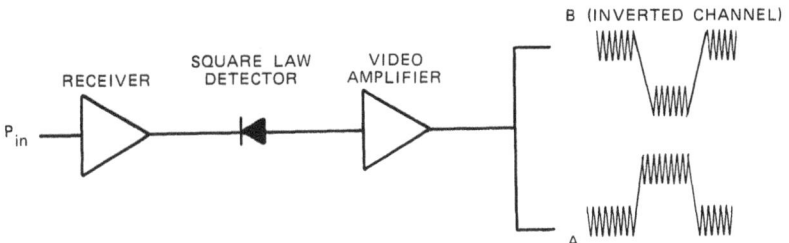

DUAL–TRACE OSCILLOSCOPE. INVERT CHANNEL B AND POSITION AS SHOWN.

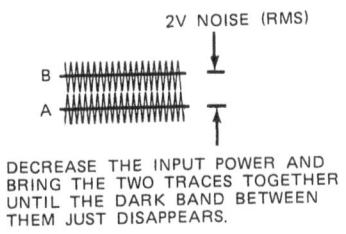

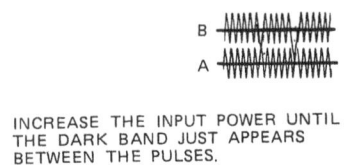

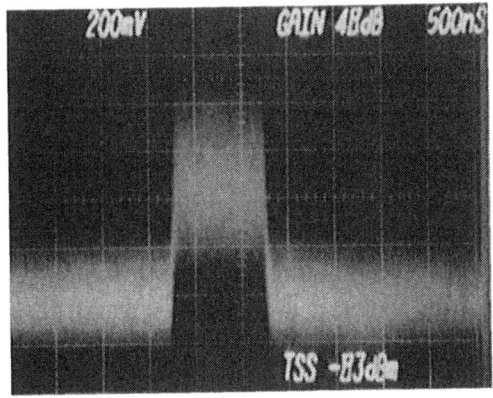

2V NOISE (RMS)

B

A

DECREASE THE INPUT POWER AND BRING THE TWO TRACES TOGETHER UNTIL THE DARK BAND BETWEEN THEM JUST DISAPPEARS.

B

A

INCREASE THE INPUT POWER UNTIL THE DARK BAND JUST APPEARS BETWEEN THE PULSES.

200mV GAIN 48dB 500nS

TSS -83dBm

INCREASE THE INPUT POWER 1 dB. THE RESULT IS THE TANGENTIAL SENSITIVITY (CHANNEL A).

FIGURE 2A-4. Tangential Sensitivity Measurement [2].

139

Tangential sensitivity depends on the noise figure* of the amplifier following the detector, the video 3-dB bandwidth, and the detector characteristics. The T_{SS} in dBm, $T_{SS}|_{dBm}$, may be given as [3,4]

$$T_{SS}\Big|_{dBm}^{G0} = -35 + 5 \, Log \, B_V + \frac{NF_V\big|_{dB}}{2} - 10 \, Log \, M \qquad (2A\text{-}1)$$

where

$$B_V = \text{video 3-dB bandwidth in MHz}$$

$$M = \text{detector constant (see Figure 2A-2)}$$

$$NF_V|_{dB} = \text{video amplifier noise figure } (10 \, Log \, F_V)$$

$$G0 = \text{receiver gain} = 0 \, dB.$$

Equation (2A-1) accurately predicts the T_{SS} for a given configuration and is an excellent comparison factor for crystal video receivers (provided B_V is specified along with T_{SS}).

The detector "effective" input signal-to-noise ratio for a T_{SS} condition is 4 dB (remember the detector is square law, Figure 1A-2 and Appendix 1B). Thus the "effective" noise appearing at the detector's input is**

$$N_D\Big|_{dBm}^{G0} = T_{SS}\Big|_{dBm}^{G0} - 4 \qquad (2A\text{-}2)$$

---

*Noise figure will be defined in dB ($NF|_{dB}$) while noise factor will be a power ratio. Thus, $NF|_{dB} \equiv 10 \, Log \, F$.

**For small gains the actual detector input noise is much less ($N_D|_{dBm} = -114 + 10 \, Log \, B_R + G_R|_{dB}$). As the receiver gain is increased, $N_D|_{dBm}$ is increased. When this noise power exceeds the "effective" noise power of Equation (2A-2), the receiver noise dominates.

and this is termed the gain-limited tangential sensitivity. If receiver gain is provided (and this gain is not sufficient to supply noise in excess of that defined by Equation (2A-2)), the detector input signal-to-noise ratio will increase 1 dB for each dB increase in gain (increasing the detector-video amplifier output signal-to-noise ratio at twice the rate of increased gain). If the receiver gain is excessively large, the noise contributed by the receiver becomes the dominant issue, and any further increase in gain results in no increase in sensitivity, but rather an equal increase of signal and noise at the input to the detector. This is termed "noise-limiting" ($T_{SS}\big|_{dBm}^{Max}$) and is the best that can be obtained for a given configuration. Figure 2A-5 illustrates the effect of receiver gain on T_{SS} (the configuration of Figure 2A-1 is used). The T_{SS} for no gain (gain-limited T_{SS}) is -41 dBm. As the gain is increased, the video noise remains constant (and T_{SS} increases 1 dB for each dB increase in gain) until the gain approaches 45 dB (Figure 2A-5d, and note the change in voltage scale factor). As the gain is increased above 45 dB (Figure 2A-5e and f), the video noise increases and little improvement in T_{SS} is gained (this is the noise-limited $T_{SS}\big|_{dBm}^{Max}$). The obvious questions now are (1) what is the noise-limited T_{SS} and (2) how much receiver gain is needed to achieve this noise-limited T_{SS}? These two questions will now be answered.

The only assumption that will be made for the following analysis is that $B_R \geq 2 B_V$. Lucas [4] presented an excellent analytic discussion on this subject; however, his equations are most difficult to use. Tsui [3] has extended Lucas' results to a more practical conclusion. The equation for $T_{SS}\big|_{dB}^{GR}$ (this is the T_{SS} for a given receiver gain, G_R) may be given as (see Appendix 2B for derivation of the equations to follow)

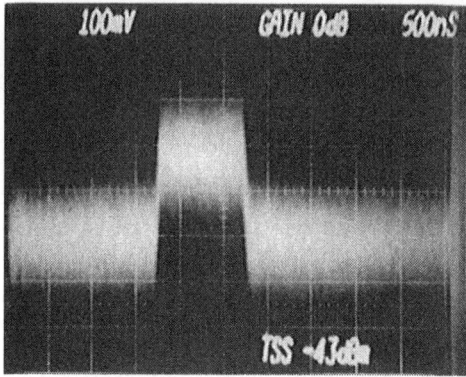

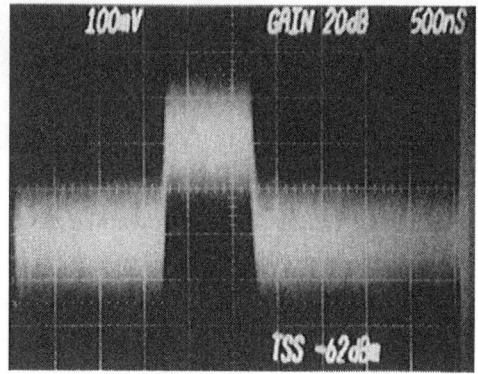

(a) Note: T_{SS} = −41 dBm not −43 dBm

(b)

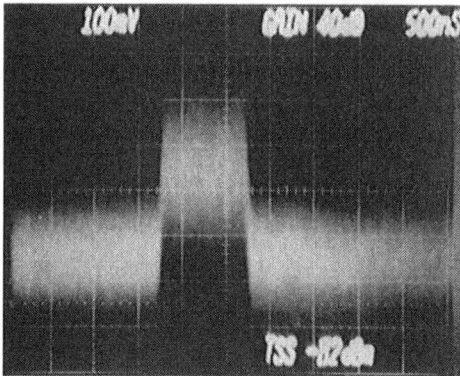

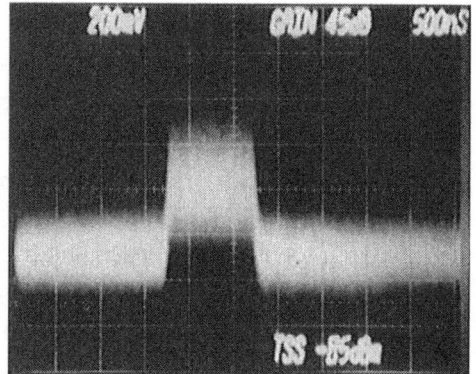

(c)

(d)

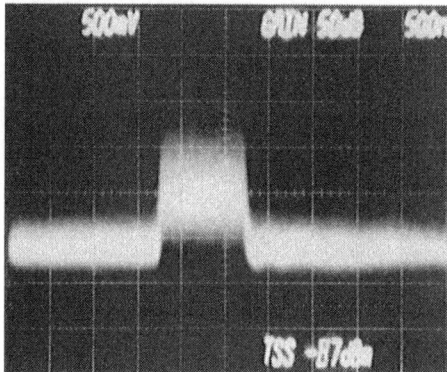

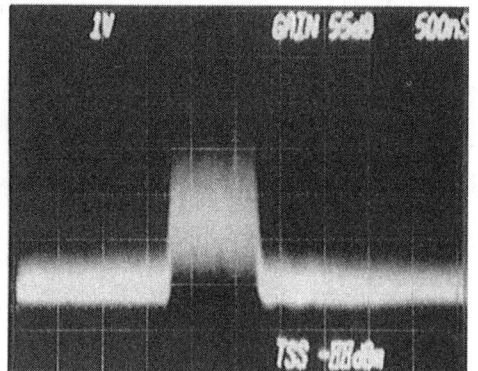

(e)

(f)

FIGURE 2A-5. Effect of Receiver Gain on T_{SS} and Output Noise,
(B_R = 40 MHz, $NF_R|_{dB}$ = 5, B_V = 7.7 MHz).

$$T_{SS}\Big|_{dBm}^{G_R} = -114 + NF_R\Big|_{dB}$$

$$+ 10 \text{ Log} \left[6.31\, B_V + 2.5 \sqrt{(2B_R B_V - B_V{}^2) + \frac{10^{\left(\frac{T_{SS}\big|_{dBm}^{G0} + 110}{10}\right)2}}{(G_R F_R)^2}} \right] \quad \text{(2A-3)}$$

where

$NF_R|_{dB}$ = receiver total noise figure in dB

B_V = video 3-dB bandwidth in MHz

B_R = receiver 3-dB bandwidth in MHz

G_R = total receiver gain (power ratio)

F_R = total receiver noise factor (power ratio, remember $NF_R|_{dB} = 10 \text{ Log } F_R$)

$T_{SS}\Big|_{dBm}^{G0}$ = T_{SS} for detector-video amplifier with no receiver gain

Two limiting cases of Equation (A-3) will now be discussed.

Case 1: $G_R F_R$ small (gain-limited sensitivity), or the right-hand term of the square root term much greater than the left. This is not obvious at first glance; however, consider the receiver parameters shown in Figure 2A-5:

B_R = 40 (MHz)

B_V = 7.7 (MHz)

$NF_R|_{dB}$ = 5 dB

$T_{SS}\Big|_{dBm}^{G0}$ = −41 dBm (see Figure 2A-5a)

143

The left term of the square root term is

$$2B_R B_V - B_V^2 = 2(40)(7.7) - (7.7)^2 = 556.7 \qquad (2A\text{-}4)$$

The right term is (if $G_R F_R = 1$)

$$\left(10^{\frac{T_{SS}\big|^{G0}_{dBm} + 110}{10}}\right)^2 = \left(10^{\frac{-41 + 110}{10}}\right)^2 = 63.1 \times 10^{12} \qquad (2A\text{-}5)$$

Thus, neglecting the left term (Equation (2A-5)) is quite valid.

Equation (2A-3) may now be written as

$$T_{SS}\bigg|^{G_R}_{dBm} = T_{SS}\bigg|^{G0}_{dBm} - G_R\bigg|_{dB} \qquad (2A\text{-}6)$$

Equation (2A-6) is most important, as it shows that increasing the gain increases the T_{SS} by the same amount. Figure 2A-5 shows that the T_{SS} with no gain is $T_{SS}\big|^{G0}_{dBm} = -41$. If 20 dB of receiver gain is added, the expected sensitivity is

$$T_{SS}\bigg|_{dBm} = -41 - 20 = -61\, dBm \qquad (2A\text{-}7)$$

and this agrees well with the measured value of -62 dBm (Figure 2A-5b).

The maximum sensitivity is obtained when $G_R F_R$ in Equation (2A-3) is increased such that the left term in the square root is dominant.

144

Case 2: $G_R F_R$ large, noise-limited sensitivity ($2B_R B_V - B_V^2$ much greater than the term to the right). Equation (2A-3) may be given as (Appendix 2B)

$$T_{SS}\Big|_{dBm}^{Max} = -114 + NF_R\Big|_{dB} + 10 \text{ Log } B_V$$

$$+ 10 \text{ Log } \left| 6.31 + 2.5 \sqrt{2 \frac{B_R}{B_V} - 1} \right| \qquad (2A\text{-}8)$$

and the maximum T_{SS} is not only dependent on the receiver noise figure, but upon the video bandwidth and the ratio of receiver to video bandwidth.

Again referring to the parameters of Figure 2A-5, we find that the maximum sensitivity expected is

$$T_{SS}\Big|_{dBm}^{Max} = -114 + 5 + 10 \text{ Log } 7.7$$

$$+ 10 \text{ Log } \left| 6.31 + 2.5 \sqrt{2\left(\frac{40}{7.7}\right) - 1} \right| \qquad (2A\text{-}9)$$

or

$$(2A\text{-}10)$$

$$T_{SS}\Big|_{dBm}^{Max} = -88.7 \, dBm$$

and this agrees quite well with the measured value of 88 dBm (Figure 2A-5f).

Later in this appendix we will discuss thresholding, and will find that increasing the receiver gain when the system is noise-limited reduces the instantaneous signal dynamic range at the detector output; thus, it is most worthwhile to find the maximum practical receiver gain. There is no pat answer as to the maximum practical receiver gain; however (as is shown in Appendix 2B), if G_R is increased (Equation 2A-3) such that

145

$$2B_R B_V - B_V^{\ 2} = \frac{\left(\dfrac{T_{SS}\Big|_{dBm}^{G0} = 110}{10} \right)^2}{(G_R F_R)^2}$$ (2A-11)

the sensitivity obtained is within 1.5 dBm of that predicted by Equation (2A-8). solving Equation (2A-11) for $G_R\Big|_{dB}^{Max}$

$$G_R\Big|_{dB}^{Max} = 110 + T_{SS}\Big|_{dBm}^{G0} - NF_R\Big|_{dB} - 10\,Log\left(\sqrt{2\,B_R B_V - B_V^{\ 2}}\right)$$ (2A-12a)

or

$$G_R\Big|_{dB}^{Max} = 110 + T_{SS}\Big|_{dBm}^{G0} - NF_R\Big|_{dB} - 10\,Log\,B_V$$

$$- 10\,Log\left(\sqrt{2\,\frac{B_R}{B_V} - 1}\right)$$ (2A-12b)

Appendix 2B illustrates this is quite close to the gain that increases the video rms noise by a factor of two over that with no receiver gain. When, in the discussions to follow, the measured value of $G_R\Big|_{dB}^{Max}$ is mentioned, it will be that receiver gain that increases the video noise by a factor of two.

The maximum predicted gain for the parameters previously discussed (Figure 2A-5) is

$$G_R\Big|_{dB}^{Max} = 110 - 41 - 5 - 10\,Log\left(\sqrt{2(40)(7.7) - (7.7)^2}\right)$$ (2A-13)

or

(2A-14)

$$G_R\Big|_{dB}^{Max} = 50.3\,dB$$

The measured value for $G_R\Big|_{dB}^{Max}$ was 49.5 dB, which gives a $T_{SS}|_{dBm}$ within 1.5 dBm of the maximum as predicted from the theory.

Several receiver configuration results are given in Table 2A-1; a plot of $T_{SS}|_{dBm}$ versus $G_R|_{dB}$ is given in Figure 2A-6. As will be seen, the predicted and measured results are quite close. Figure 2A-7 is a summary of the theory presented thus far. Applying this theory to the more useful area of thresholding will be presented shortly; however, it is worthwhile to discuss how the material presented relates to the work of others.

It will be seen from Figure 2A-5e that the noise riding on the pulse is greater than the baseline noise. This condition is predicted by Lucas [4], and the situation becomes more pronounced as B_R/B_V decreases. Figure 2A-8 shows $T_{SS}\big|_{dBm}^{Max}$ for a receiver with $B_R/B_V = 80$, and as can be seen, the noise riding on the pulse is indeed nearly equal to the standing noise. This situation is inconsequential, however, as Lucas shows that $T_{SS}\big|_{dBm}^{Max}$ is fairly independent (0.5 dBm or so) of this "excess" noise.

Ayer [5] was one of the first to present a coherent discussion of detector-video amplifiers preceded by RF (or IF) amplification. His work is a pioneering effort, and he defines an "effective bandwidth," B_e, as

$$B_e = \sqrt{2B_R B_V - B_V^2} \tag{2A-15}$$

This is based on the assumption that B_R/B_V is quite large, which is often not the case. Many subsequent authors neglected this assumption, and one can have serious errors in the predicted $T_{SS}\big|_{dBm}^{Max}$. Referring to Equation (2A-12), we see that our predicted maximum gain is a function of Ayer's B_e; however, $T_{SS}\big|_{dBm}^{Max}$ (Equation 2A-8) is not a direct function of B_e. Appendix 2B shows that, if $B_R/B_V > 50$ (for $B_R/B_V = 50$ the error using B_e is 0.98 dBm), $T_{SS}\big|_{dBm}^{Max}$ may be approximated as

$$T_{SS}\bigg|_{dBm}^{Max} \simeq -110 + NF_R\bigg|_{dB} + 10 \text{ Log } B_e \text{ (for } B_R/B_e > 50) \tag{2A-16}$$

TABLE 2A-1. Comparison of Predicted and Measured Results (PW = 1μsec).

B_R, MHz	B_V, MHz	$NF_R|_{dB}$	$TSS|_{dBm}^{G_0}$	$TSS|_{dBm}^{Max}$		$G_R|_{dB}^{Max}$		B_R/B_V
				Measured[a]	Equation (2A-8)	Measured[b]	Equation (2A-12)	
40	7.7	5	-41	-88	-88.7	51	50.27	5.2
120	7.7	6	-41	-86	-87.1	48	47.8	15.6
40	0.5	5	-44c	-95	-95.0	53	53.0	80
120	0.5	6	-44c	-92	-93.0	50	50.6	240

[a] Measured at $G_R|_{dB}^{Max}$ (measured value).

[b] This is the receiver gain needed to double the output video rms noise (see Appendix 2B).

[c] Equation (2A-1) is a function of 5 Log B_V; thus,if B_V is decreased from 7.7 (MHz) to 0.5 (MHz) the $TSS|_{dB}^{G_0}$ should increase by 5 Log 7.7/0.5 = 5.9 dB. This assumes the M and $NF_V|_{dB}$ remain constant. It was found that as B_V was decreased, $NF_V|_{dB}$ increased by 2.5 dB, thus accounting for the 3-dBm, rather than 5.9-dBm, increase in TSS.

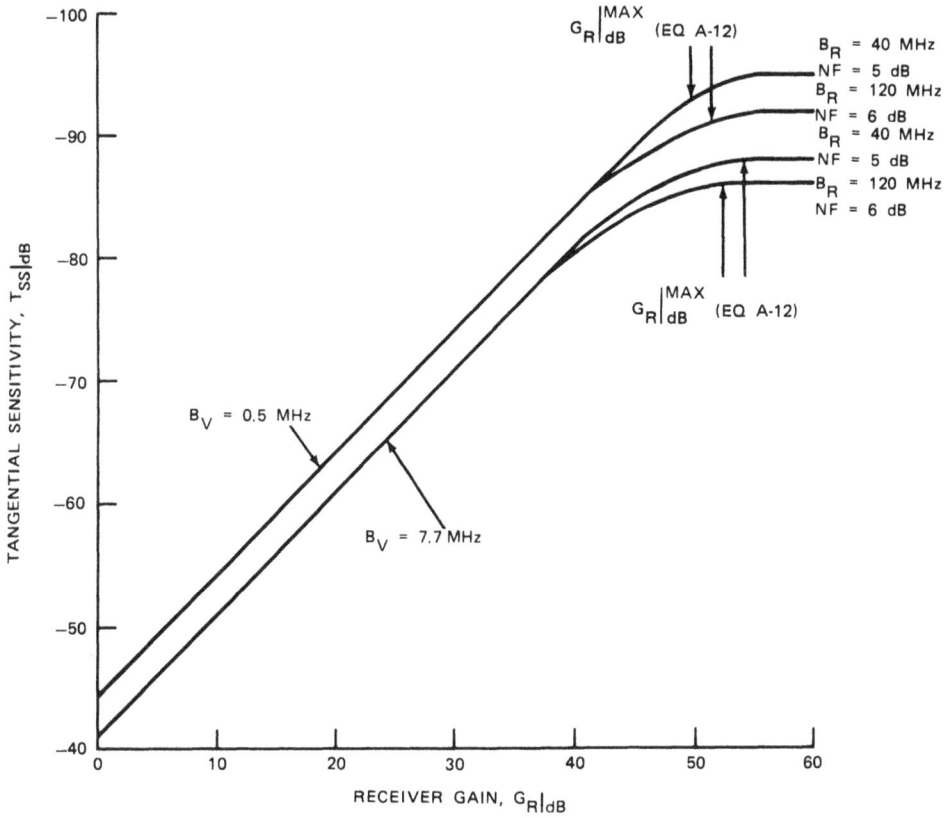

FIGURE 2A-6. Tangential Sensitivity Versus Gain.

which is the same as predicted by Ayer (provided one adds $+4$ dB to his equation to account for the 4-dB signal-to-noise ratio at the detector input for a T_{SS} condition).

149

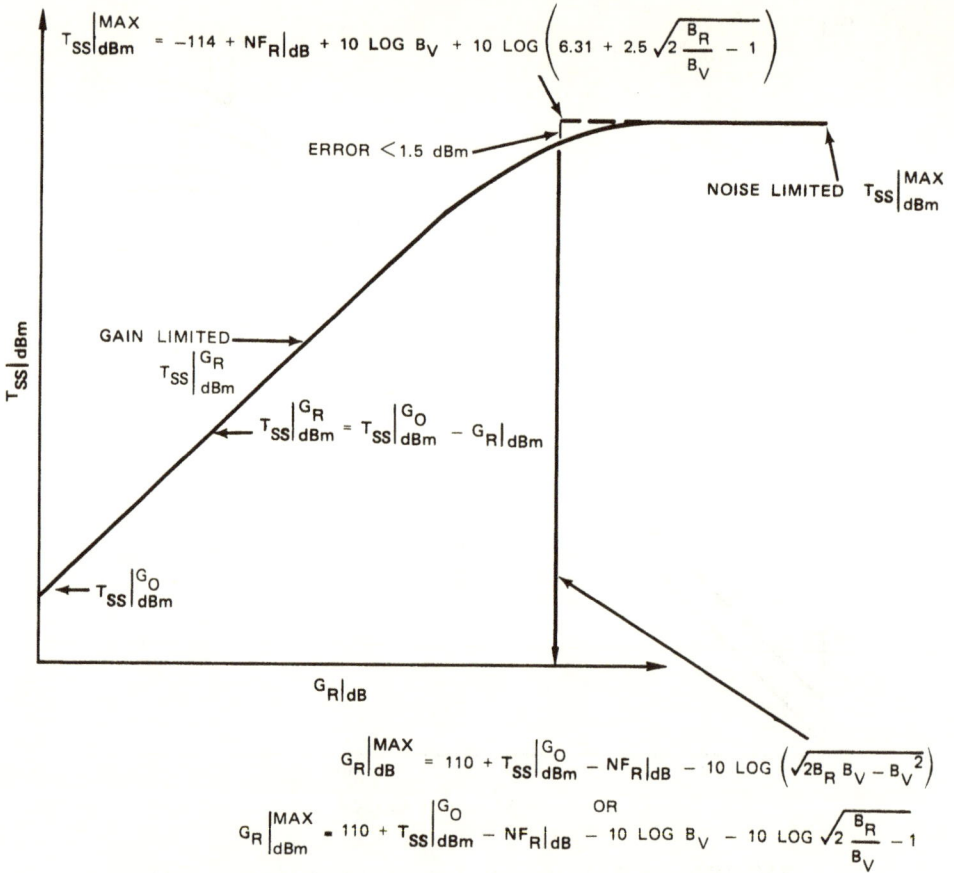

$$T_{SS}\Big|_{dBm}^{MAX} = -114 + NF_R\Big|_{dB} + 10\ LOG\ B_V + 10\ LOG\left(6.31 + 2.5\sqrt{2\frac{B_R}{B_V} - 1}\right)$$

ERROR < 1.5 dBm

NOISE LIMITED $T_{SS}\Big|_{dBm}^{MAX}$

GAIN LIMITED

$T_{SS}\Big|_{dBm}^{G_R}$

$$T_{SS}\Big|_{dBm}^{G_R} = T_{SS}\Big|_{dBm}^{G_O} - G_R\Big|_{dBm}$$

$T_{SS}\Big|_{dBm}^{G_O}$

$T_{SS}\Big|_{dBm}$

$G_R\Big|_{dB}$

$$G_R\Big|_{dB}^{MAX} = 110 + T_{SS}\Big|_{dBm}^{G_O} - NF_R\Big|_{dB} - 10\ LOG\left(\sqrt{2B_R B_V - B_V^2}\right)$$

OR

$$G_R\Big|_{dBm}^{MAX} = 110 + T_{SS}\Big|_{dBm}^{G_O} - NF_R\Big|_{dB} - 10\ LOG\ B_V - 10\ LOG\sqrt{2\frac{B_R}{B_V} - 1}$$

FIGURE 2A-7. Summary of T_{SS} and G_{Max} Equations.

Basic Thresholding

$T_{SS}|_{dBm}$ is a reasonable parameter to compare receiver-detector-video amplifiers with similar B_R and B_V; however, the real world of electronic warfare and radar requires a threshold to be exceeded to indicate the presence of a signal. The effect of B_R, B_V, and G_R on this threshold will now be presented.

150

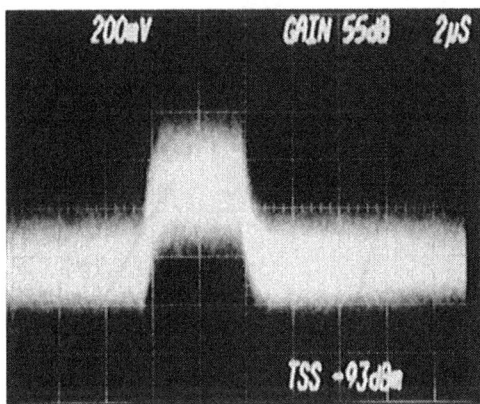

FIGURE 2A-8. $T_{SS}\Big|_{dBm}^{Max}$ for $B_R/B_V = 80$.

Figure 2A-1 illustrates the general configuration to be presented. The detector-video amplifier drives a threshold comparator as shown. With no signal present, if signal threshold voltage, $V_T|_S$, is too low, noise due to the receiver-detector-video amplifier will trigger the threshold, giving a large noise-count termed false-alarm rate (FAR). As $V_T|_S$ is increased, this false signal count decreases. We will be concerned with the threshold voltage that gives a FAR count of one per second (FAR = 1). Figure 2A-9 illustrates the FAR count versus $V_T|_S$ as the receiver gain of Figure 2A-1 is varied (this is the same configuration used for Table 2A-1). Several observations concerning Figure 2A-9 should be noted:

1. For a given gain, FAR is a strong function of $V_T|_S$.

2. The FAR curves are a function of $G_R|_{dB}$. This is to be expected at large gains $(G_R\Big|_{dB}^{Max})$ since the receiver is contributing noise.

151

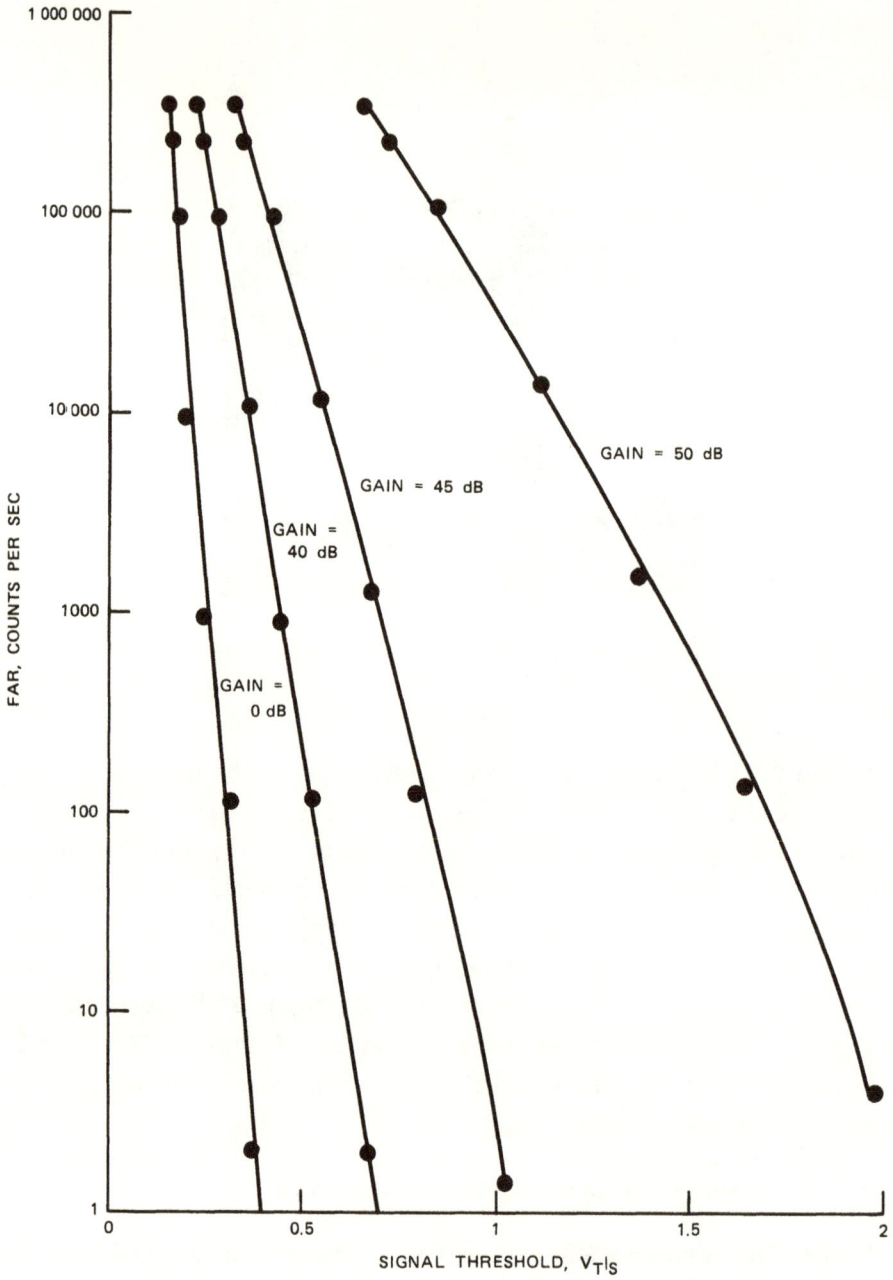

FIGURE 2A-9. False Alarm Rate Versus $V_T|S$
(B_R = 120 MHz, B_V = 7.7 MHz, $NF_R|dB$ = 6).

3. For receiver gains less than 45 dB ($G_R \big|_{dB}^{Max}$ = 47.8, from Table 2A-1), the FAR (log scale) curves versus $V_T|_S$ are fairly linear.

4. For receiver gain in excess of $G_R \big|_{dB}^{Max}$, the threshold voltage necessary to give FAR = 1 becomes excessive.

Item 4 is most important in receiver systems employing variable gain. If $V_T|_S$ is set for 0.4 volt (FAR = 1 for G_R (dB) = 0 dB), the FAR will be 100,000 if the gain is increased to 45 dB. This is obviously an unreasonable situation. One obvious solution would be to set $V_T|_S \simeq 1$ volt, which gives FAR = 1 for a gain of 45 dB. However, as the gain is decreased (decreasing the noise), only 0.4 volt of $V_T|_S$ is needed. Obviously, the FAR for 0-dB gain and $V_T|_S = 1$ volt is quite low (one count every 15 minutes or so); however, the input signal to the comparator must increase to 1 volt (rather than 0.4 volt) to trigger the threshold. Thus a loss in detector output dynamic range occurs. This loss in output dynamic range, referred to the input of the square law detector, may be given as

$$D_R \bigg|_{dBm}^{Loss} = 10 \, Log \frac{V_T \big|_S^A}{V_T \big|_S^B} \tag{2A-17}$$

where

$V_T \big|_S^A$ = signal threshold for gain A (FAR = 1)

$V_T \big|_S^B$ = signal threshold for gain B (FAR = 1)

As an example (refer to Figure 2A-9), the threshold for a gain of 50 dB is 2 volts, and for a gain of 0 dB is 0.4 volt. The loss in dynamic range is

$$D_R \bigg|_{dBm}^{Loss} = 10 \, Log \frac{2}{0.4} = 7 \, dB \tag{2A-18}$$

or if the threshold is fixed at 2 volts and the gain is decreased to 0 dB, the receiver input must be increased 7 dBm to exceed the 2-volt threshold.

153

Reasonable Criterion for Signal Sensitivity

You are probably wondering at this point what criterion should be used to determine if a signal is present: A threshold crossing for all pulses? 50%? Or what? We will use the 80% probability of detection ($P_d|80\%$) criterion--80% of the signal pulse repetition frequency (PRF) will trigger the signal threshold (for a PRF of 2 kHz, a count of 1.6 kHz is the 80% $P_d|80\%$). Figure 2A-10 shows the 80% probability of detection sensitivity ($P_d|_{dBm}^{80\%}$), as a function of receiver gain, for a FAR = 1 (with no signal, $V_T|s$ is increased until FAR = 1; then the signal is increased until an 80% signal count is obtained). A plot of T_{SS} is also given, and it should be noted that $P_d|_{dBm}^{80\%}$ is between 1 and 2 dBm lower than $T_{SS}|_{dBm}$. The loss in instantaneous dynamic range, $D_R|_{dBm}^{Loss}$ (Equation (2A-17)) and the rms video noise are also shown. the important point here is that for receiver gains larger than $G_R|_{dBm}^{Max}$, the video noise increases, necessitating an increase in $V_T|s$ (to keep FAR = 1), thus increasing $D_R|_{dBm}^{Loss}$. A reasonable estimate in the loss in $D_R|_{dBm}^{Loss}$ for gains in excess of $G_R|_{dBm}^{Max}$ is

$$DR\bigg|_{dBm}^{Loss} \simeq G_R\bigg|_{dBm} - G_R\bigg|_{dBm}^{Max} + 6 \qquad (2A\text{-}19)$$

Figures A-11 through A-16 illustrate FAR versus $V_T|s$ and $P_d|_{dBm}^{80\%}$ for the receiver parameters given in Table 2A-1. It is interesting to note that $P_d|_{dBm}^{80\%}$ for large receiver gains is within 1 dBm of the measured $T_{SS}|_{dBm}^{Max}$. At lower gains (gain-limited sensitivity), the difference between $T_{SS}|_{dBm}^{80\%}$ is between 2 and 3 dB.

Table 2A-2 is a summary of the data taken for the receivers presented. Several important observations should be noted:

1. $P_d|_{dBm}^{80\%}$ measured at $G_R|_{dB}^{Max}$ is within 1 dBm of the maximum possible.

2. $P_d|_{dBm}^{80\%}$ maximum is within 2 dB of the maximum tangential sensitivity.

3. The loss in dynamic range is approximately 6 dB for $G_R|_{dB}^{Max}$.

Thus, from a practical standpoint, the 80% probability of detection sensitivity (FAR = 1) may be approximated as

$$P_d \bigg|_{dBm}^{80\%} \simeq T_{SS} \bigg|_{dBm}^{Max} - 2 \qquad \text{(2A-20)}$$

and the maximum gain for this sensitivity is equal to the maximum gain obtained from Equation (2A-12).

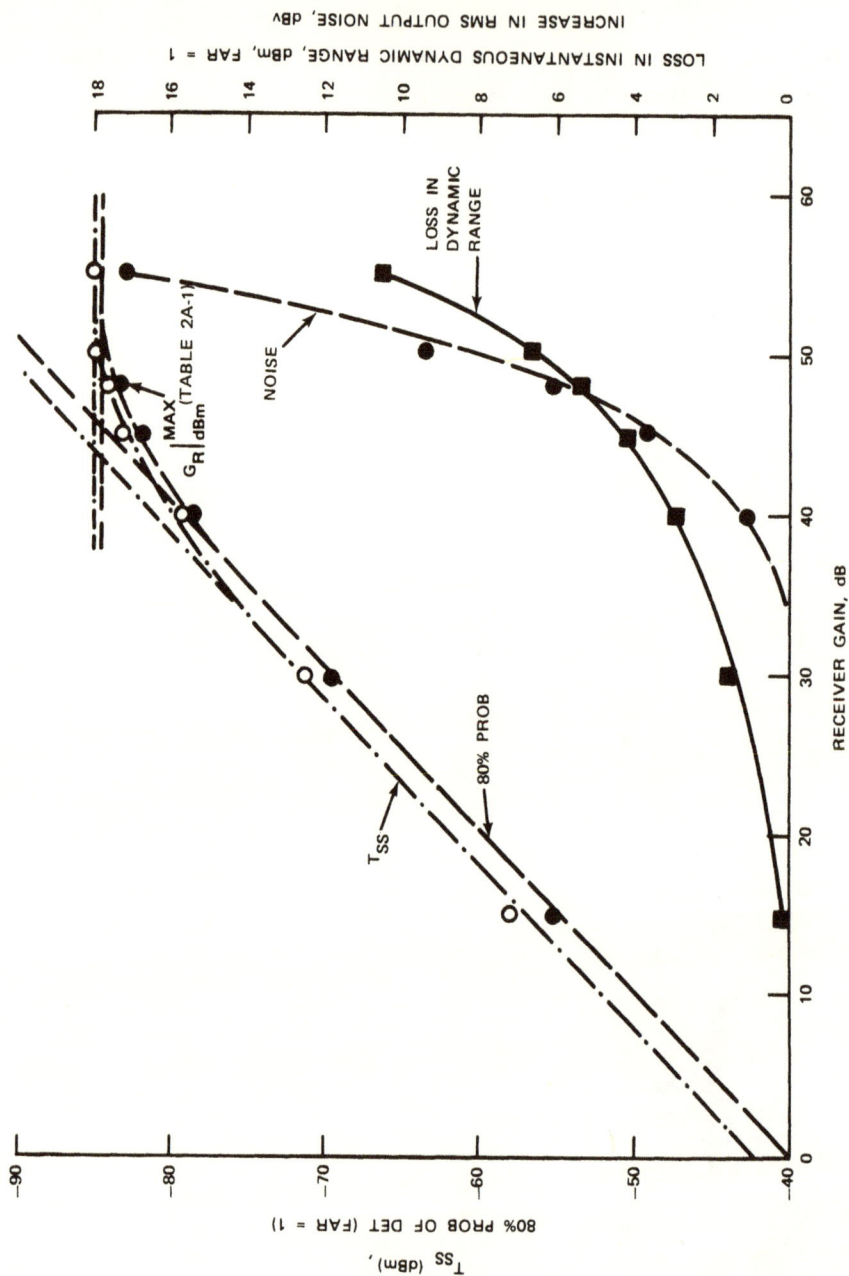

FIGURE 2A-10. Sensitivity Versus Receiver Gain (B_R = 120 MHz, B_V - 7.7 MHz, $NF_R|dB$ = 6).

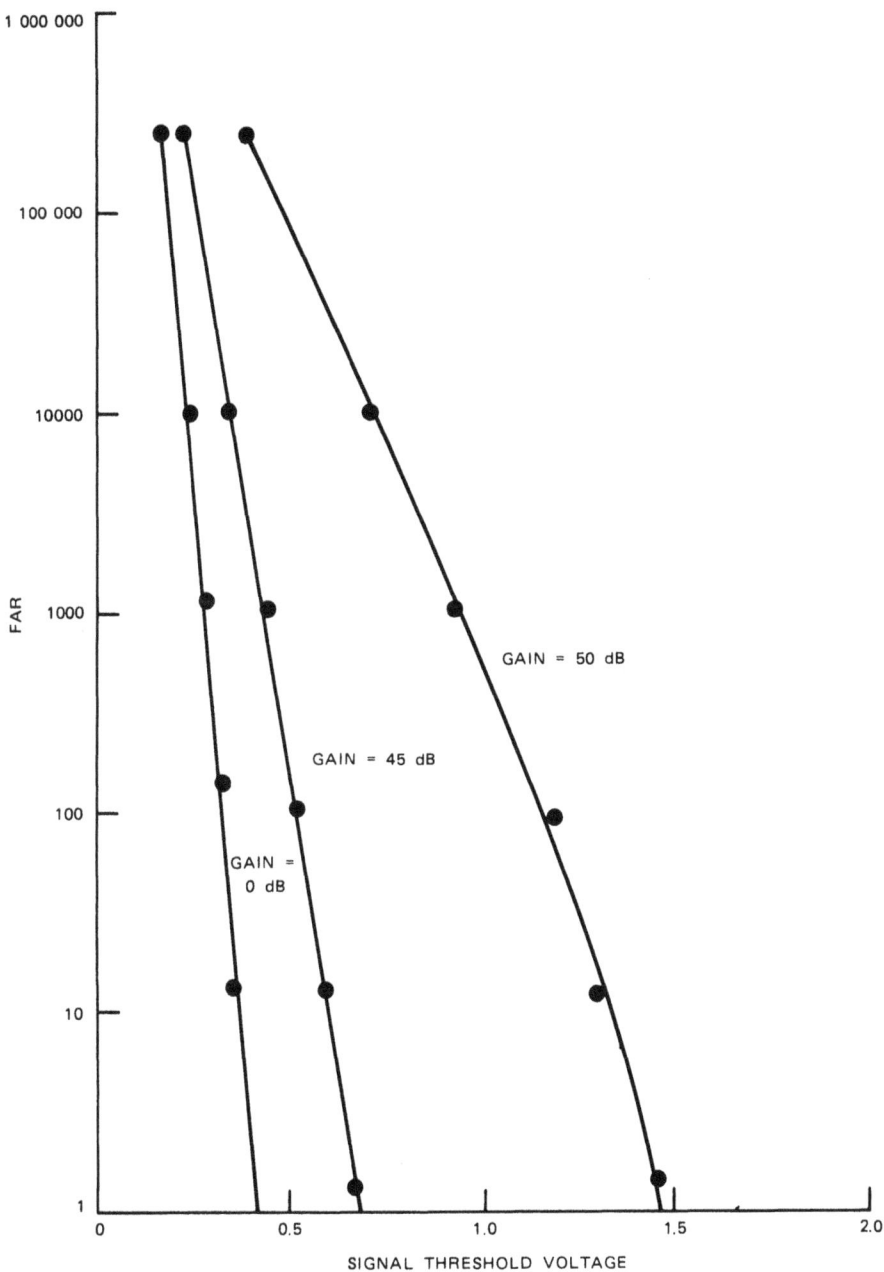

FIGURE 2A-11. False Alarm Rate Versus $V_T|s$
($B_R = 40$ MHz, $B_V = 7.7$ MHz, $NF_R|dB = 5$).

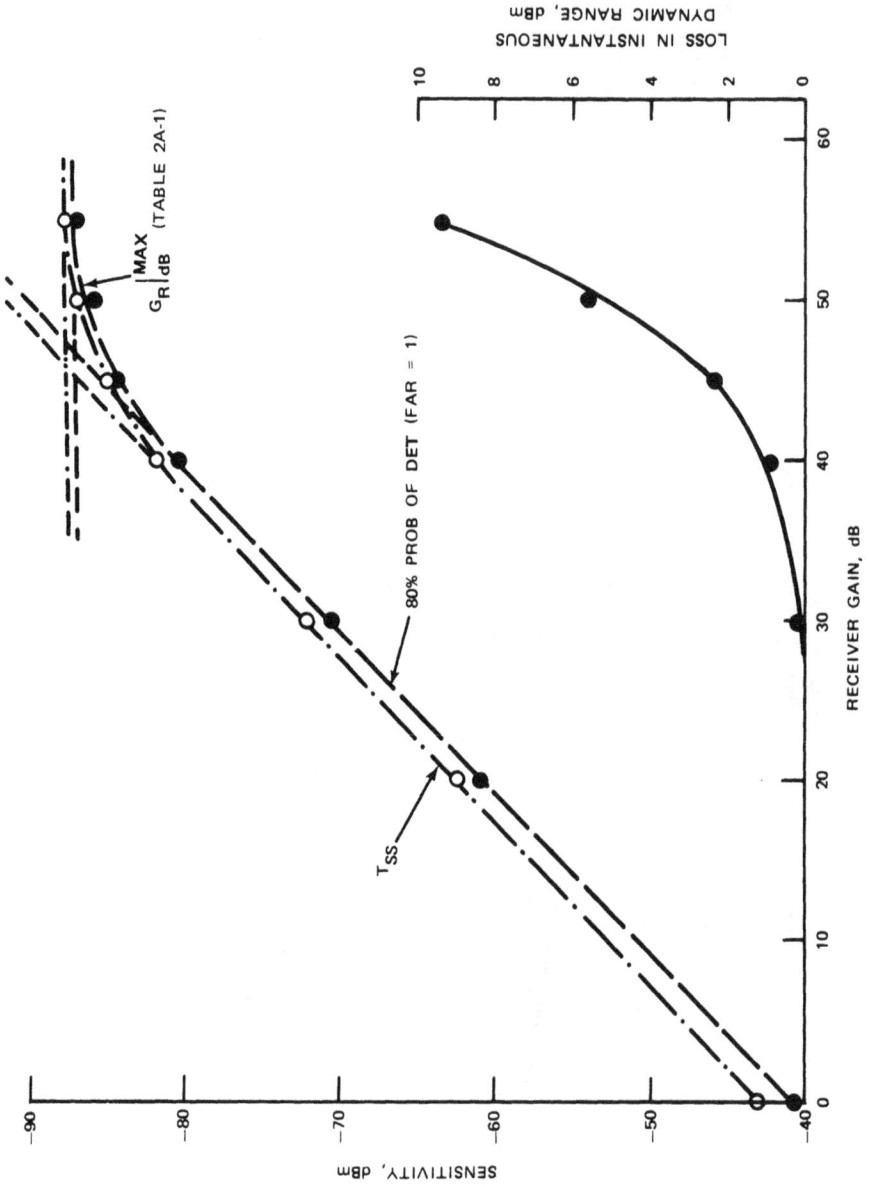

FIGURE 2A-12. Sensitivity Versus Receiver Gain
(B_R = 40 MHz, B_V - 7.7 MHz, $NF_R|dB$ = 5).

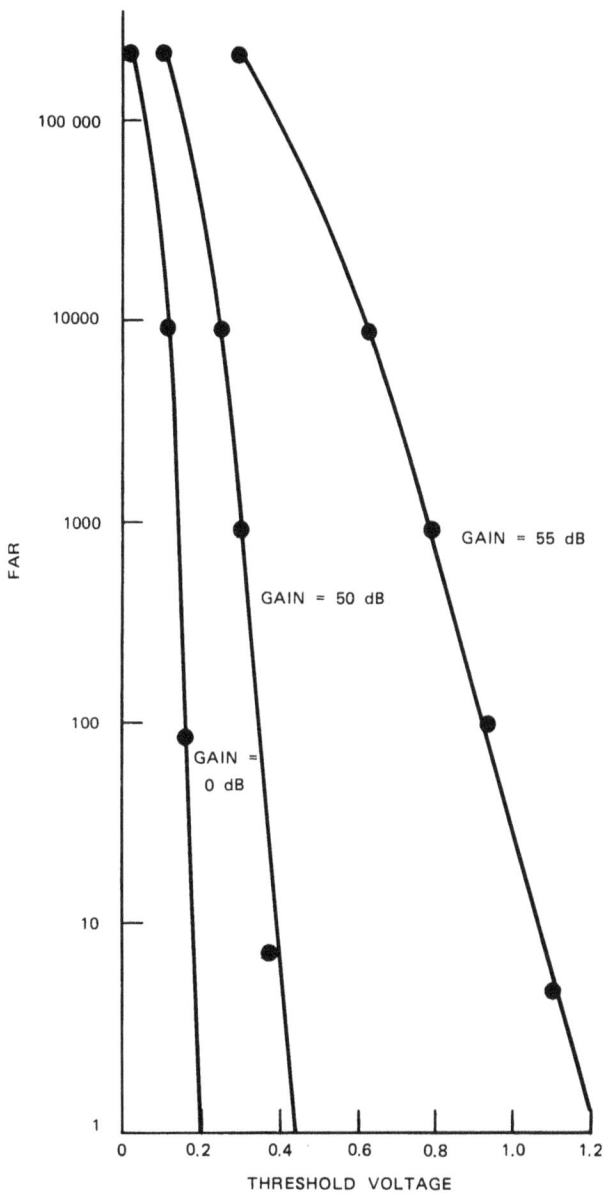

FIGURE 2A-13. False Alarm Rate Versus $V_T|S$
(B_R = 120 MHz, B_V = 0.5 MHz, $NF_R|dB$ = 6).

159

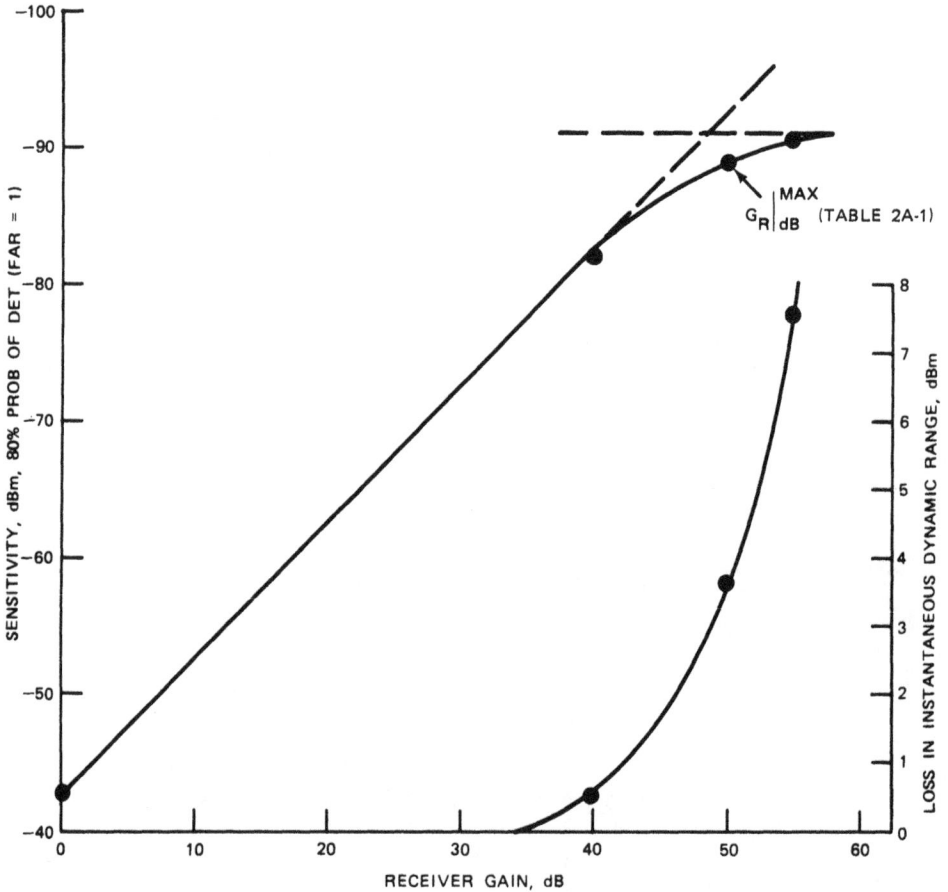

FIGURE 2A-14. Sensitivity Versus Receiver Gain
(B_R = 120 MHz, B_V = 0.5 MHz, $NF_R|_{dB}$ = 6).

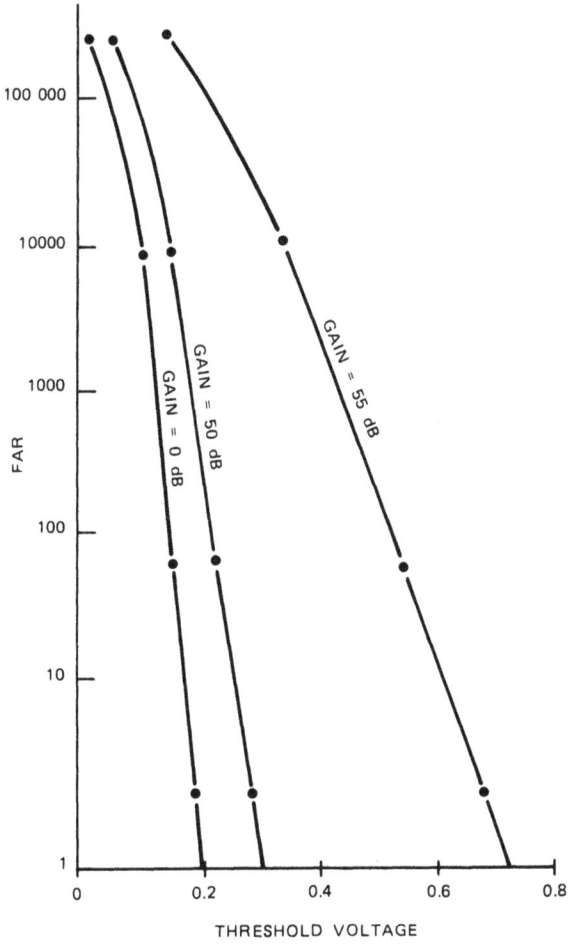

FIGURE 2A-15. FAR Versus $V_T|_S$ (B_R = 40 MHz, B_V = 0.5 MHz, $NF_R|_{dB}$ = 5).

161

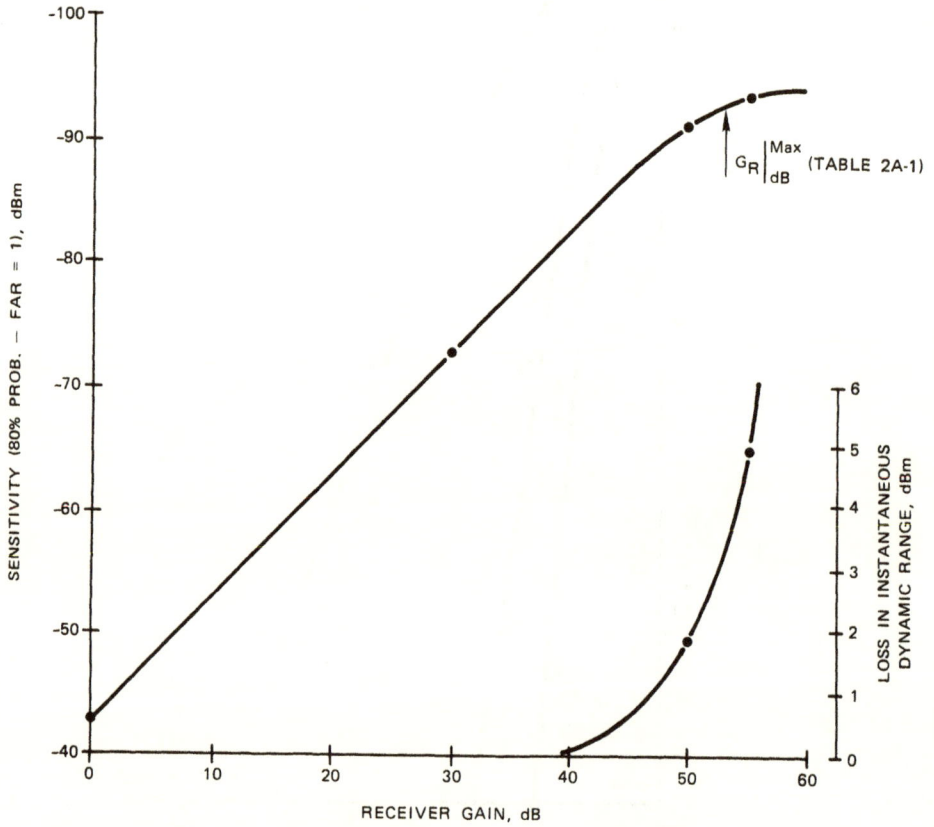

FIGURE 2A-16. Sensitivity Versus Receiver Gain
(B_R = 40 MHz, B_V = 0.5 MHz, $NF_R|_{dB}$ = 5).

162

Appendix 2B

DERIVATIONS OF $T_{SS}\Big|_{dBm}^{GR}$, $T_{SS}\Big|_{dBm}^{Max}$, AND $G_R\Big|_{dB}^{Max}$ EQUATIONS

Tsui solves Lucas' T_{SS} equation as (for $B_R > 2 B_V$)

$$T_{SS}\Big|_{dBm} = -114 + NF_R\Big|_{dB}$$

$$+ 10 \operatorname{Log} \left[6.31 B_V + 2.5 \sqrt{(2B_R B_V - B_V^2) + \frac{AB_V}{(G_R F_R)^2}} \right] \qquad (2B\text{-}1)$$

where

$$A = \frac{4 F_V R_V}{KTK_{SL}^2} \times 10^{-6} \qquad (2B\text{-}2)$$

and

$R_V =$ detector video resistance

$T =$ temperature in degrees Kelvin

$K_{SL} =$ detectors square law K in mV/mW

$F_V =$ video amplifier noise factor

$K =$ Boltzmann's constant (1.38×10^{-23} J/°K)

If $G_R F_R = 1$, AB_V is found to be $\gg 2B_R B_V - B_V^2$, and Equation 2B-1 may be solved for A

$$A = \frac{\left(10^{\frac{T_{SS}\Big|_{dBm}^{G0} + 110}{10}} \right)^2}{B_V} \qquad (2B\text{-}3)$$

Limiting Case 1, $AB_V/(G_RF_R)^2 \gg (2B_RB_V - B_V^2)$*

$$T_{SS}\Big|_{dBm} = -114 + NF_R\Big|_{dBm} + 10 \text{ Log}\left(2.5\sqrt{\frac{AB_V}{(G_RF_R)^2}}\right) \qquad (2B\text{-}4)$$

Substituting Equation (2B-3) for A

$$T_{SS}\Big|_{dBm} = -110 + \left(10 \text{ Log}\sqrt{\left(10^{\frac{T_{SS}\big|_{dBm} + 110}{10}}\right)^2}\right) - G_R\Big|_{dB} \qquad (2B\text{-}5)$$

or

$$T_{SS} = T_{SS}\Big|_{dBm}^{G0} - G_R\Big|_{dB} \qquad (2B\text{-}6)$$

Limiting Case 2, $(2B_RB_V - B_V^2) \gg AB_V/(G_RF_R)^2$

$$T_{SS}\Big|_{dBm}^{Max} = -114 + NF_R\Big|_{dB} + 10 \text{ Log}\left[6.31\, B_V + 2.5\sqrt{2B_RB_V - B_V^2}\right] \qquad (2B\text{-}7)$$

or

$$T_{SS}\Big|_{dBm}^{Max} = -114 + NF_R\Big|_{dB} + 10 \text{ Log}\,(B_V)\left(6.31 + 2.5\sqrt{2\left(\frac{B_R}{B_V}\right)-1}\right) \qquad (2B\text{-}8)$$

Many authors neglect the 6.31 B_V term in the equation. This assumption, valid only if $B_R \gg B_V$, yields a T_{SS} (Equation (2B-7)) of

$$T_{SS}\Big|_{dBm} = -114 + NF_R\Big|_{dB} + 10 \text{ Log}\,(2.5)\left(\sqrt{2B_RB_V - B_V^2}\right) \qquad (2B\text{-}9a)$$

or

$$T_{SS}\Big|_{dBm}^{Max} = -110 + NF_R\Big|_{dB} + 10 \text{ Log}\sqrt{2B_RB_V - B_V^2} \qquad (2B\text{-}9b)$$

---

*It should be noted that $\sqrt{AB_V/(G_RG_T)^2}$ is much greater than 6.31 B_V.

164

The term $\sqrt{2B_R B_V - B_V^2}$ has been termed the "effective bandwidth" and, if it can be used, simplifies T_{SS} calculations since a single effective bandwidth can be assigned to a given receiver. Comparing Equations (2B-8) and (2B-9a), we see that the only error is to the right of the 10 Log term. Equation (2B-9a) may now be given as

$$T_{SS}\Big|_{dBm}^{Max} = -114 + NF_R\Big|_{dB} + 10 \text{ Log }(2.5\,B_V)\left(\sqrt{2\,\frac{B_R}{B_V} - 1}\,\right) \qquad (2B\text{-}10)$$

or

$$T_{SS}\Big|_{dBm}^{Max} = -114 + NF_R\Big|_{dB} + 10 \text{ Log } B_V + 10 \text{ Log}\left(2.5\,\sqrt{2\,\frac{B_R}{B_V} - 1}\,\right) \qquad (2B\text{-}11)$$

Thus the error from the actual predicted $T_{SS}|_{dBm}$ (Equation (2A-8)) and approximate $T_{SS}|_{dBm}^{Max}$ (Equation (2B-10)) is

$$\text{Error}\Big|_{dB} = \text{Equation B-10} - \text{Equation B-11} \qquad (2B\text{-}12)$$

or

$$\text{Error}\Big|_{dB} = 10 \text{ Log}\left(6.31 + 2.5\,\sqrt{2\,\frac{B_R}{B_V} - 1}\,\right) - 10 \text{ Log}\left(2.5\,\sqrt{2\,\frac{B_R}{B_V} - 1}\,\right) \qquad (2B\text{-}13)$$

Table 2B-1 gives this error for various B_R/B_V ratios and, as can be seen, $B_R/B_V > 50$ has an error within 1 dB. Thus, using Equation (2A-9b) to predict $T_{SS}|_{Max}^{dB}$ can give larger than expected predicted $T_{SS}|_{dBm}^{Max}$.

TABLE 2B-1. Error in Assuming $B_R/B_V \gg 1$.

B_R/B_V	$10 \ \text{Log} \left(6.31 + 2.5 \sqrt{2 \dfrac{B_R}{B_V} - 1} \right)$	$10 \ \text{Log} \left(2.5 \sqrt{2 \dfrac{B_R}{B_V} - 1} \right)$	$\text{Error} \big|_{dB}$
2	10.27	6.37	3.9
5	11.4	8.75	2.65
10	12.36	10.37	1.98
50	14.94	13.96	0.98
100	16.19	15.47	0.71

Determining the maximum gain necessary to provide $T_{SS} \big|_{Max}^{dBm}$ is not, as stated earlier, a straightforward matter; however, solving for the gain necessary to give Equation (2B-1):

$$\frac{AB_V}{(G_R F_R)^2} = 2B_R B_V - B_V^{\ 2} \tag{2B-14}$$

does give quite reasonable results. Substituting Equation (2B-3) into Equation (2B-14) and solving for G_R

$$G_R \bigg|_{dB}^{Max} = T_{SS} \bigg|_{dBm}^{G0} + 110 - NF_R \bigg|_{dB} \tag{2B-15}$$

$$- 10 \ \text{Log} \ \sqrt{2B_R B_V - B_V^{\ 2}}$$

The $T_{SS}|_{dB}$ for this condition may be easily solved as

$$T_{SS}\Big|_{dBm}^{G_R|_{dB}^{Max}} = -114 + NF_R\Big|_{dB} \qquad (2B\text{-}16)$$

$$+ \ 10 \ Log \ (B_V)\left(6.32 + 3.54 \ \sqrt{2 \ \frac{B_R}{B_V} - 1}\right)$$

The error in $T_{SS}|_{dBm}$ predicted (Equation (2B-16)), and the maximum $T_{SS}|_{dBm}$ expected (Equation (2B-8)), is summarized in Table 2B-2. As can be seen, the $T_{SS}|_{dB}$ predicted using $G_R|_{dB}^{Max}$ is slightly lower than expected; however, this error is at worst -1.5 dB (for large B_R/B_V ratios). It should be noted that increasing $G_R|_{dB}$ 1 dB above our calculated value does not increase $T_{SS}|_{dB}$ by 1 dB, as the two terms in Equation (2B-14) are still nearly equal. To actually obtain the $T_{SS}|_{dBm}^{Max}$ predicted by Equation (2B-8), $G_R|_{dB}$ must be increased 5 dB or so above $G_R|_{dB}^{Max}$ given in Equation (2B-15). This increase in gain increases the video noise and causes serious thresholding problems. It has been found experimentally that the value for $G_R|_{dB}^{Max}$ of Equation (2B-15) approximately doubles the rms noise out of the video amplifier (with respect to the noise with no receiver gain). Thus $G_R|_{dB}^{Max}$ may be regarded, in principle at least, as that gain needed to produce a receiver noise power equal to the detector-video amplifier noise power at the detector's input.

TABLE 2B-2. Error in $T_{SS}|_{dBm}$ Using $G_R|_{dB}^{Max}$.

B_R/B_V	$10 \log(6.31 + 2.5\sqrt{2B_R/B_V - 1})$	$10 \log(6.31 + 2.5\sqrt{2B_R/B_V - 1})$	$Error|_{dB}$
2	10.27	6.37	3.9
5	11.40	8.75	2.65
10	12.36	10.37	1.98
50	14.94	13.96	0.98
100	16.19	15.47	0.71
$\geqslant 100$	...	...	-1.5

Appendix 2C

DERIVATION OF $T_{SS}|_{dBm}$, $T_{SS}|_{dBm}^{Max}$, AND $G_R|_{dB}^{Max}$
FOR $B_V \leq B_R \leq 2B_V$

Tsui [3] presents Lucas' [4] results as

$$T_{SS}\Big|_{dBm} = -114 + NF_R\Big|_{dB}$$

$$+ 10 \log \left[3.15B_R + 2.5 \sqrt{2B_R B_V - B_V^2} + \frac{AB_V}{(G_R F_R)^2}\right] \qquad (2C-1)$$

The only difference between Equation (2C-1) and Equation (2B-1), is the first term after the Log ($6.31B_V$ for Equation (2B-1) and $3.15B_V$ for Equation (2C-1)). The A term is defined in Equation (2B-3). The limiting cases may now be given as

$$\text{Limiting Case 1:} \quad \frac{AB_V}{(G_R F_R)^2} \gg 2B_R B_V - B_V^2$$

$$T_{SS}\Big|_{dBm} = T_{SS}\Big|_{dBm}^{G0} - G_R\Big|_{dB} \qquad (2C-2)$$

which is the same as Equation (2B-6) and is to be expected.

$$\text{Limiting Case 2:} \quad (2B_R B_V - B_V^2) \gg \frac{AB_V}{(G_R F_R)^2}$$

$$T_{SS}\Big|_{dBm}^{Max} = -114 + NF_R\Big|_{dB} + 10 \log(B_V)$$

$$+ 10 \log \left[3.15 + \sqrt{2 \frac{B_R}{B_V} - 1}\right] \qquad (2C-3)$$

169

This equation gives a T_{SS} some 1.5 to 2.3 dBm better than the T_{SS} if Equation (2B-8) is used.

The receiver gain necessary to give $T_{SS}|_{dBm}^{Max}$ is found in accordance with Equation (2B-14)

$$\frac{AB_V V}{(G_R F_R)^2} = 2B_R B_V - B_V^2 \tag{2C-4}$$

and Equation (2B-15) is obtained

$$G_R\Big|_{dBm}^{Max} = T_{SS}\Big|_{dBm}^{G0} + 110 - NF_R\Big|_{dB} - 10 \log \sqrt{2B_R B_V - B_V^2} \tag{2C-5a}$$

or

$$G_R\Big|_{dBm}^{Max} = T_{SS}\Big|_{dBm}^{G0} + 110 - NF_R\Big|_{dB} - 10 \log B_V$$

$$- 10 \log \sqrt{2\frac{B_R}{B_V} - 1} \tag{2C-5b}$$

It has been found experimentally that for $B_V > B_R$, replacing $10 \log B_V$ with $10 \log B_R$ and neglecting the $\sqrt{2 B_R/B_V - 1}$ terms in Equations (2C-3) and (2C-5b) gives quite reasonable results.

170

Nomenclature

$A_V/_{ST}$ signal threshold noise amplifier gain

B_e effective bandwidth

B_R receiver Bandwidth

B_V video bandwidth

D noise integrator duty cycle

$DR\big|_{dB}^{Loss}$ receiver instantaneous dynamic range loss

f_{3dB} noise riding threshold loop 3-dB frequency response

FAR false alarm rate

FAR_N normalized FAR count

F_V video amplifier noise factor

$G0$ receiver gain $= 0$ dB

$G_R\big|_{dB}$ receiver gain (dB)

H noise loop transfer function

LG loop gain

M detector diode constant

N normalized noise count

n noise loop contrast

$ND\big|_{dBm}^{G0}$ effective detector noise input for $G_R = 0$

$NF_R\big|_{dB}$ receiver noise figure (dB)

o/s one shot

$P_D|_{dBm}$ detector input (dBm)
$P_d|_{dBm}$ signal input for 80% probability of detection
$P_{in}|_{dBm}$ receiver input power (dBm)
PRF signal pulse repetition frequency
PW signal pulse width

S_{DAC} DAC scale factor (volt/bit)
S_F noise loop scale factor (FAR/volt)

T temperature in degrees Kelvin
T_C counter enable time
T_I integrator enable (update) time
$T_{SS}|_{dBm}$ tangential signal sensitivity
$T_{SS}|_{dBm}^{G0}$ T_{SS} for no receiver (receiver gain = 0)
$T_{SS}|_{dBm}^{Max}$ maximum possible T_{SS} obtainable

V_{DAC} digital to analog converter effective output voltage
V_ε error voltage
V_{Ref} reference voltage
$V_{T/N}$ noise threshold voltage
$V_{T/S}$ signal threshold voltage

Δt noise loop closure time

Bibliography

Ayer, W. E. *Characteristics of Crystal-Video Receivers Employing RF Preamplification.* Stanford Electronic Laboratories, Stanford University, Palo Alto, CA, 20 September 1956. (Tech Report No. 150-3, publication UNCLASSIFIED.)

Frohmaier, J. H. "Noise Performance of a Three-Stage Microwave Receiver." *Correspondence of Electronic Technology* (England), June 1960, pp. 245-46.

Lucas, W. J., and E. R. A. Beck. "Noise Performance of a Three-Stage Microwave Receiver." *Correspondence of Electronic Technology* (England), February 1962, pp. 76-78.

Klipper, H. "Sensitivity of Crystal Video Receiver With RF Preamplification." *Microwave Journal,* Vol. 8, 1965, pp. 85-92.

Lucas, W. J. "Tangential Sensitivity of a Detector Video System With RF Preamplification." *Proceedings of the IEEE,* Vol. 113, No. 8 (August 1966), pp. 1321-30.

Harp, J. C. "What Does Receiver Sensitivity Mean?" *MSN,* July 1978, pp. 54-63.

Tsui, J. "Tangential Sensitivity of EW Receivers." *Microwave Journal,* October 1981, pp. 99-102.

Tsui, J. and Show, R. "Sensitivity of EW Receivers." *Microwave Journal,* November 1982, pp. 115-120.

Tsui, J. "False Alarm Measurements on Receivers." *Microwave Journal*, September 1984, pp. 213-217.

Naval Weapons Center. *Determining Maximum Sensitivity and Optimum Maximum Gain for Detector-Video Amplifiers with RF Preamplification*, by R. S. Hughes. China Lake, CA, NWC, March 1985. (NWC TM 5357, publication UNCLASSIFIED.)

Hughes, R. S. "Determining Maximum Sensitivity and Optimum Maximum Gain for Detector-Video Amplifiers with RF Preamplification." *Microwave Journal*, November 1985, pp. 165-178.

Tsui, J. *Microwave Receivers with Electronic Warfare Applications*. John Wiley and Sons, New York, 1986.

Sakaie, Y. "Probability of False Alarm and Threshold of Square Law Detector-Logarithmic Amplifier for Narrow Band-Limited Gaussian Noise." *IEEE Transactions on Circuits and Systems*, Vol. CAS-33, No. 1, January 1986, pp. 59-65.

Tsui, J. "An Approximation Equation for EW Receiver Sensitivity." *Microwave Journal*, September 1986, pp. 214-218.

Farina, A. and Studer, F. A. "A Review of CFAR Detection Techniques in Radar Systems." *Microwave Journal*, September 1986, pp. 115-128.

Lipsky, S. *Microwave Passive Direction Finding*. John Wiley and Sons, New York, 1987.

Remmell, J. A. and Hughes, R. S. "Converting T_{SS} to Detection Receiver Sensitivity." *Microwaves and RF*, December 1987, pp. 112-125.

Chapter 3

RANGE-TRACKING LOOPS

Many radar systems (both airborne and land or ship based) employ range trackers as an integral part of their pulse processing. This chapter will present the basics of range tracker operation, and a detailed analysis and design procedures (with examples) for the heart of range tracking: the range tracking integrators.

Target acquisition and reacquisition techniques will not be covered, and the discussions to follow assume that range tracker lock-on has occurred (as will be shown, the range track loop is, in essence, a phase-locked loop and the acquisition and lock-on concepts are similar).

Figure 3-1 illustrates a simplified active pulse radar. The transmit-receive (T/R) switch allows the transmitted signal to be radiated by the antenna. The T/R switch then connects the antenna to a mixer and the received signal is mixed to the IF frequency, amplified, detected, and fed to the pulse processing circuitry. The range-tracking loop enables the AGC loop, angle tracker, and target detection and acquisition circuitry only when an expected return pulse is present.

The range tracker must be capable of following any changes in closing velocity between the radar and target, and if the target return should fade, the range enable gates should still be generated until the target reappears or reacquisition is initiated. The loop bandwidth should also be as low as possible to limit the effectiveness of countermeasure techniques used against range-tracking loops.

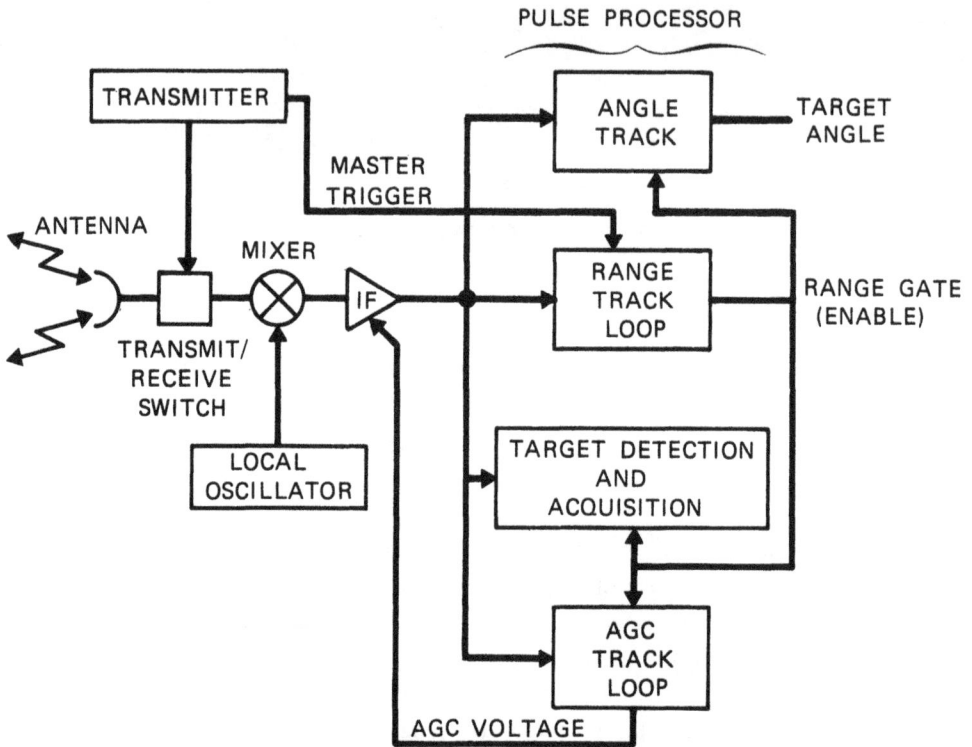

FIGURE 3-1. Basic Active Pulse Radar.

Figure 3-2 illustrates a basic airborne radar-target configuration. The radar transmits at time T_0, and the transmitted pulse reaches the target at time T_1. The pulse is then reflected and reaches the radar at time T_2. It is obvious that $T_1 = T_2$. Thus, the range to the target, R (assuming the pulse travels at the speed of light), is

$$R = \frac{C \Delta t}{2} \text{ (ft)} \qquad (3\text{-}1)$$

where

c = speed of light (984 ft/μsec)

$\Delta t = T_1 + T_2$ (total time of pulse travel in μsec)

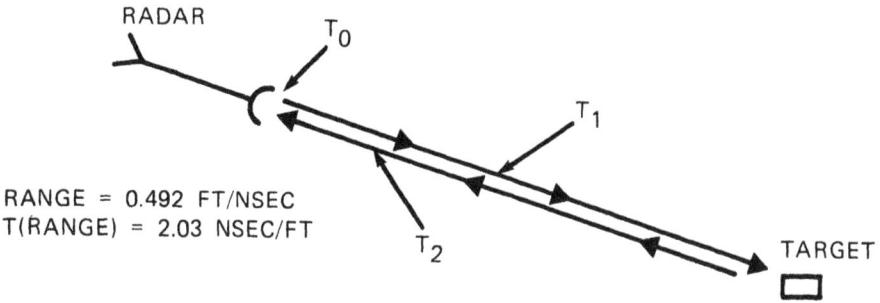

FIGURE 3-2. Typical Radar Target Configuration.

Figure 3-3 illustrates the basic range-tracking loop configuration (it should again be emphasized that target acquisition has occurred). At time T_0, the T/R switch allows signal transmission and also starts the linear range ramp. When the range ramp voltage equals the range integrator voltage, ER (Figure 3-4a), the comparator output changes state, triggering the range enable and early-late gate. The early-late gates enable the range discriminator, the output of which is continually "zeroed" by the feedback action of the loop (i.e., the range gate is driven such that the target return is centered in the range gate). The loop filter determines the dynamic loop performance.

The configuration shown in Figure 3-3 does not lend itself to straightforward analysis. Appendix 3A discusses the range-tracking loop from a phase-locked loop perspective, and the range-tracking loop can be presented (for analysis purposes) as illustrated in Figure 3-5. We will first do a general analysis of the functional range-tracking loop (without specifying the loop filter) and then replace $F(S)$ with two filters that are in general use.

177

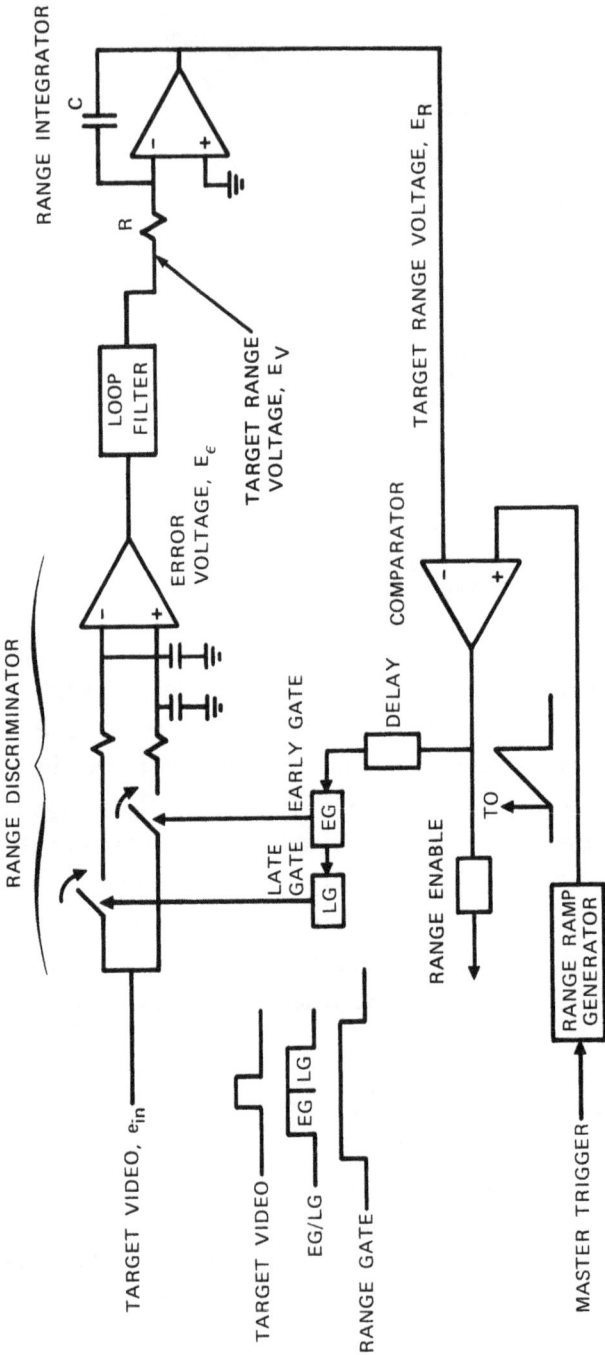

FIGURE 3-3. Basic Range Track Loop (Type II, Zero Velocity Error).

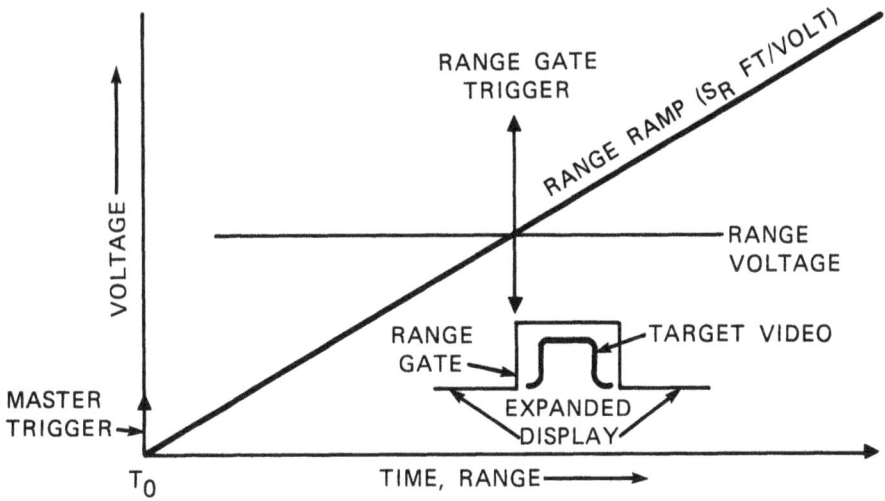

(a) Range ramp-range voltage timing.

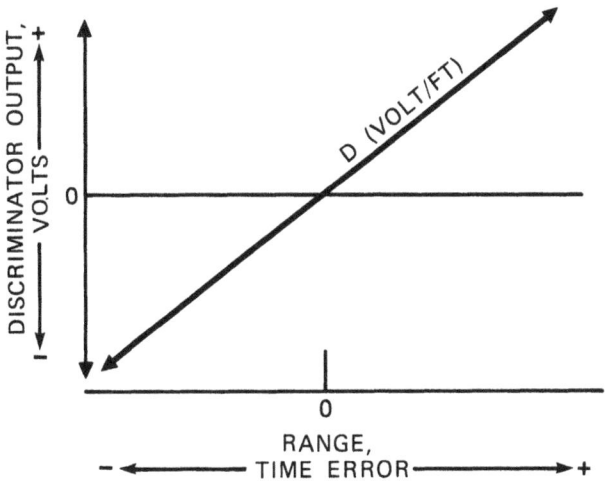

(b) Discriminator volt-error curve.

FIGURE 3-4. Range Ramp and Discriminator Timing.

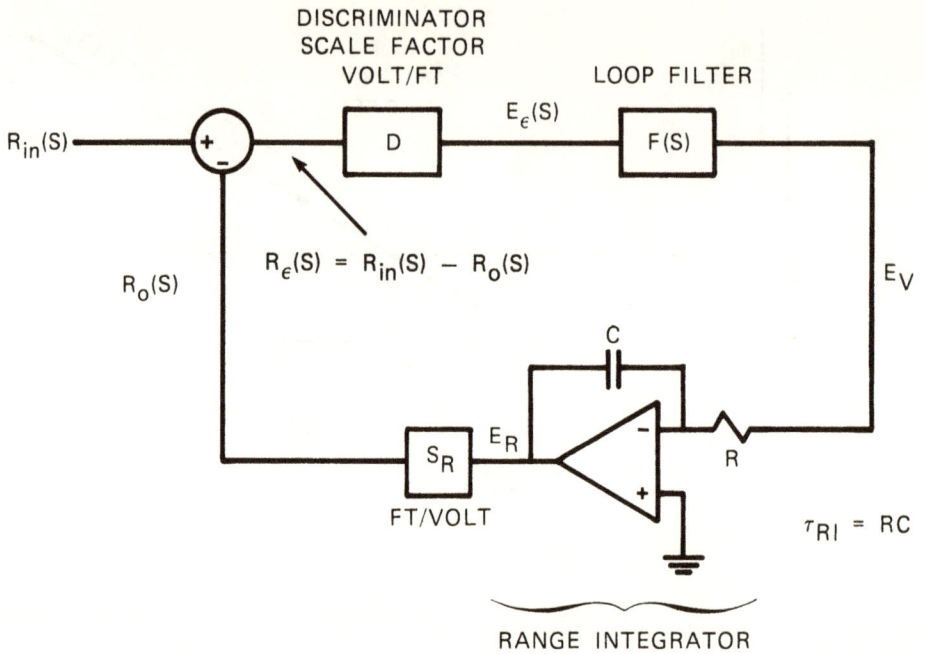

FIGURE 3-5. Functional Range-Tracking Loop.

The general loop gain equation may be given as (Figure 3-5)

$$LG(S) = (D) F(S) \left(\frac{1}{RCS} \right) S_R \qquad (3\text{-}2)$$

or

$$LG(S) = \frac{DS_R F(S)}{\tau_{RI} S} \qquad (3\text{-}3)$$

The loop transfer function, $[R_o(S)]/[R_{in}(S)]$, may be given as

$$H(S) = \frac{R_o(S)}{R_{in}(S)} = \frac{LG(S)}{1 + LG(S)} \qquad (3\text{-}4)$$

Substituting Equation (3-3) into (3-4),

$$H(S) = \frac{\dfrac{DS_R \, F(S)}{t_{RI}}}{S + \dfrac{DS_R \, F(S)}{t_{RI}}} \tag{3-5}$$

The normalized range error is

$$\frac{R_{in}(S) - R_o(S)}{R_{in}(S)} = \frac{R_\varepsilon(S)}{R_{in}(S)} \tag{3-6}$$

or

$$1 - \frac{R_o(S)}{R_{in}(S)} = \frac{R_\varepsilon(S)}{R_{in}(S)} = 1 - H(S) \tag{3-7}$$

Substituting Equation (3-5) into (3-7),

$$\frac{R_\varepsilon(S)}{R_{in}(S)} = \frac{S}{S + \dfrac{DS_R}{t_{RI}} F(S)} \tag{3-8}$$

Solving Equation (3-8) for the range error, we have

$$R_\varepsilon(S) = R_{in}(S) \frac{S}{S + \dfrac{DS_R}{t_{RI}} F(S)} \tag{3-9}$$

The importance of Equation (3-9) lies in the fact that the loop dynamics to step input changes in range ($\Delta R_{in}S$), velocity ($\Delta Vel/S^2$), and acceleration ($\Delta Accel/S^3$) may be obtained. The final range error can then be found by the *final value theorum*:

$$\underset{t \to \infty}{\text{Lim}} R_\varepsilon(t) = \underset{S \to 0}{\text{Lim}} S[R_\varepsilon(S)] \tag{3-10}$$

The first loop filter that will be covered (and the one generally used) is illustrated in Figure 3-6. This filter ensures a zero final range error to a velocity step as will be shown.

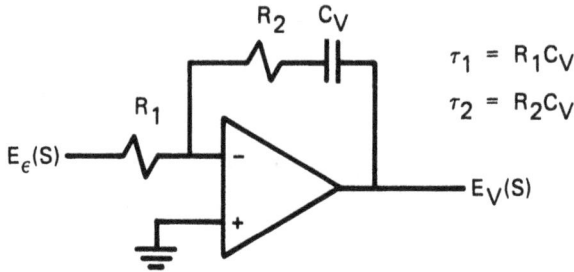

FIGURE 3-6. Zero Velocity Error (Type II) Loop Filter.

The transfer function, assuming a large operational amplifier gain, may be given as*

$$F(S) = \frac{E_v(S)}{E_\epsilon(S)} = \frac{1 + S \tau_2}{S \tau_1} \tag{3-11}$$

Substituting Equation (3-11) into (3-3), the loop gain becomes

$$LG(S) = \frac{DS_R}{\tau_{RI} \tau_1} \left| \frac{1 + S \tau_2}{S^2} \right| \tag{3-12}$$

---

*The inverting sign for the filter (and the range integrator) will be omitted as it is easily accounted for in the final design.

Letting

$$K = \frac{D S_R}{\tau_{RI}} \qquad (3\text{-}13)$$

$$LG(S) = \frac{K}{S^2} \left[\frac{1 + S \tau_2}{\tau_1} \right] \qquad (3\text{-}14)$$

The loop transfer function, H(S) (Equation (3-4)), may now be given as

$$H(S) = \frac{\dfrac{K}{S} \left[\dfrac{1 + S \tau_2}{S \tau_1} \right]}{1 + \dfrac{K}{S} \left[\dfrac{1 + S \tau_2}{S \tau_1} \right]} \qquad (3\text{-}15)$$

or, upon simplifying,

$$H(S) = \frac{\dfrac{K}{\tau_1} + \dfrac{K \tau_2}{\tau_1} S}{S^2 + \dfrac{K \tau_2}{\tau_1} S + \dfrac{K}{\tau_1}} \qquad (3\text{-}16)$$

This is the classic equation of a second-order system*

$$H(S) = \frac{\omega_n^2 + 2 \zeta \omega_n S}{S^2 + 2 \zeta \omega_n + \omega_n^2} \qquad (3\text{-}17)$$

---

* Order is defined as the highest degree of the denominator of the characteristic equation. Since we have an S^2 in the denominator of Equation (3-17), this is a second-order system.

183

where ω_n is the loop natural frequency and ζ is the damping factor. From Equations (3-16) and (3-17),

$$\omega_n = \sqrt{K/\tau_1} = \sqrt{\frac{DS_R}{\tau_{RI}\tau_1}} \tag{3-18}$$

$$\zeta = \frac{\tau_2}{2}\sqrt{K/\tau_1} = \frac{\tau_2}{2}\sqrt{\frac{DS_R}{\tau_{RI}\tau_1}} \tag{3-19}$$

Equations (3-18) and (3-19) illustrate that the natural frequency, ω_n, can be made independent of ζ by ensuring

$$\frac{\tau_2}{2}\omega_n = \text{constant} \tag{3-20}$$

The importance of Equation (3-20) will be discussed momentarily.

In addition to being a second-order system, Equation (3-20) is also a Type II loop.* Thus from classic servo theory we expect the final value for the range error, R_ε, found in Table 3-1.

TABLE 3-1. Final Range Error for a Second-Order, Type II System.

Input	Final Error ($R_{\varepsilon(\text{final})}$)
Step range	0
Step velocity	0
Step acceleration	Constant

---

*Loop type is defined as the number of poles of the loop gain (Equation (3-15)) located at the origin.

Substituting the filter Equation (3-11) into the range error Equation (3-9) and solving for the range error as a function of time yields complicated results. Fortunately, these equations have been solved and graphed for phase-locked loops [1 and 2], and, since the configuration of Figure 3-5 (and the previous discussion) is based on phase-locked loop theory, they are equally valid for range-tracking loops.

Figures 3-7 through 3-9 are plots of normalized range error for a step in range, velocity, and acceleration, respectively. As expected, in Figures 3-7 and 3-8, the final range error is zero. In Figure 3-9 the final range error is nonzero and approaches

$$R_\varepsilon(\text{final}) \simeq \frac{\Delta \text{Accel}}{\omega_n^2} \qquad (3\text{-}21)$$

Figure 3-10 summarizes the general equations for a second-order, Type II range-tracking loop for easy reference.

Second-Order, Type II, Range-Tracker Design Procedure

The following paragraphs describe the procedure for designing a second-order, Type II, range tracker. Figure 3-11 illustrates the dual bandwidth (fast for velocity acquisition, slow for acceleration tracking), second-order, Type II, range-tracking loop that will be discussed.

A. Knowing maximum range in feet ($R_{max}(\text{ft})$), determine range ramp sweep width (T_{SW}) and scale factor, S_v.

$$T_{SW} = 1.016 \times 10^9 \left(\frac{\text{sec}}{\text{ft}} \right) R_{max}(\text{ft}) \text{ sec} \qquad (3\text{-}22)$$

185

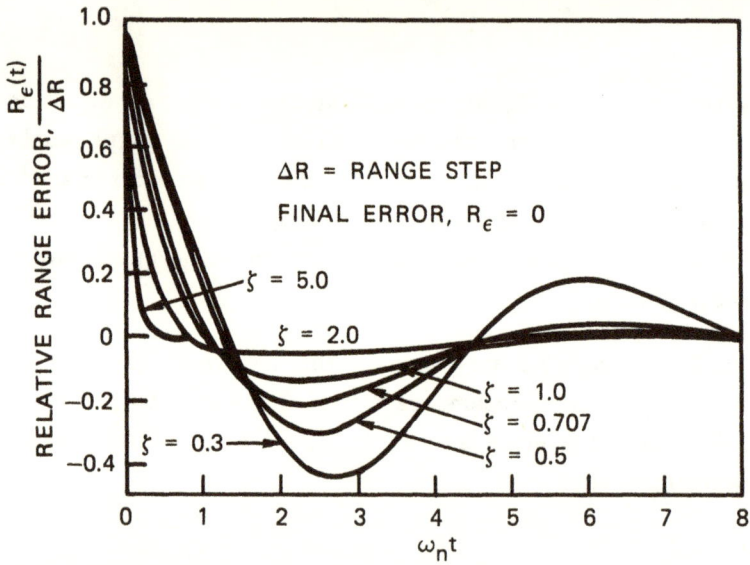

FIGURE 3-7. Range Error, $R_\epsilon(t)$, due to a Step in Range, ΔR (Second-Order, Type II Loop).

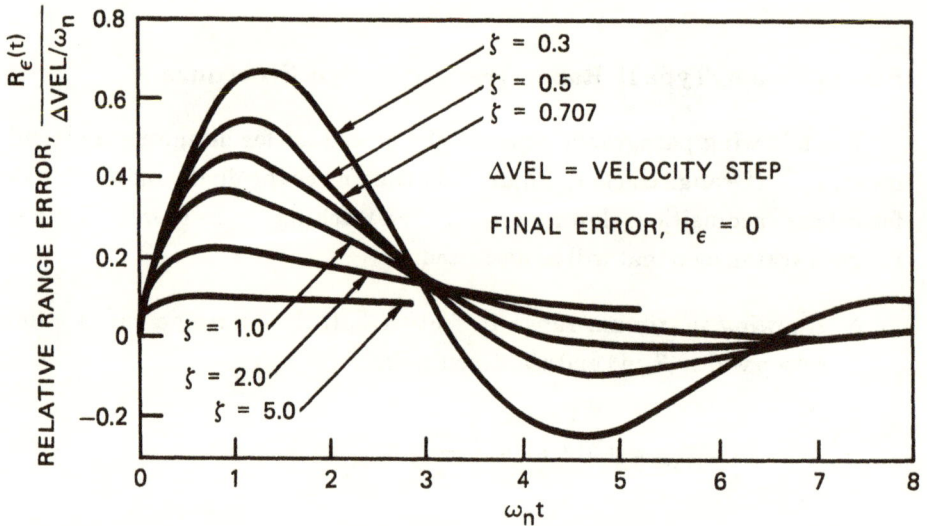

FIGURE 3-8. Range Error, $R_\epsilon(t)$, due to a Step in Velocity, ΔVel (Second-Order, Type II Loop).

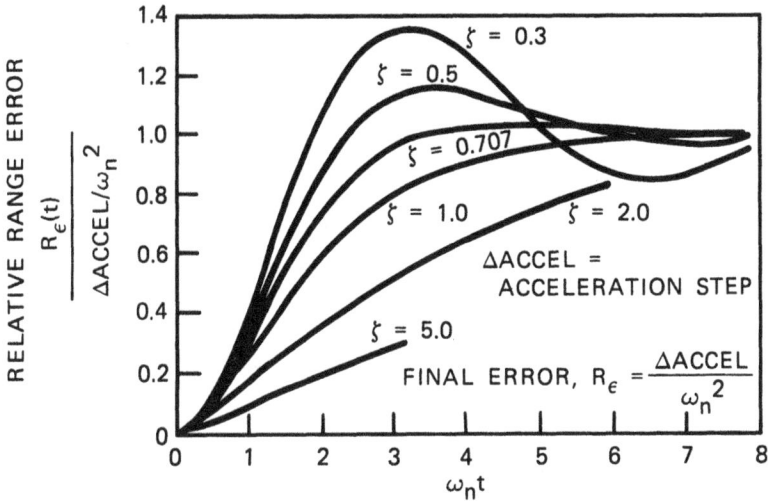

FIGURE 3-9. Range Error, $R_\epsilon(t)$, due to a Step in Acceleration, ΔAccel (Second-Order, Type II Loop).

The maximum value for the sweep voltage (E_{SM}) is only limited to the supply voltages available.

$$S_v = \frac{E_{SM}}{T_{SW}} \text{ V/sec} \qquad (3\text{-}23)$$

B. Knowing the range ramp scale factor, calculate S_R (ft/volt)

$$S_R = \frac{R_{max}(ft)}{E_{SM}} = \frac{R_{max}(ft)}{S_v(volt/sec)\,T_{SW}(sec)} \text{ ft/V} \qquad (3\text{-}24)$$

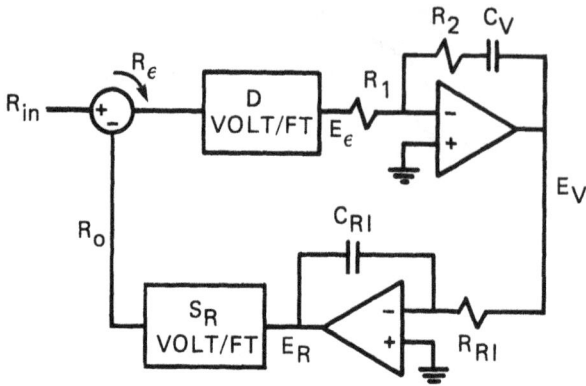

$$\tau_1 = R_1 C_V$$
$$\tau_2 = R_2 C_V$$
$$\tau_{RI} = R_{RI} C_{RI}$$

$$LG(S) = \frac{DS_R}{\tau_{RI} S^2} \left[\frac{1 + S\tau_2}{\tau_1} \right]$$

$$\frac{R_o(S)}{R_{in}(S)} = \frac{\omega_n^2 + 2\zeta\omega_n S}{S^2 + 2\zeta\omega_n + \omega_n^2}$$

$$\omega_n = \sqrt{\frac{DS_R}{\tau_{RI}\tau_1}} \quad \zeta = \frac{\tau_2}{2}\,\omega_n$$

$$\text{NOISE BANDWIDTH, } BW_N = \frac{\omega_n}{2}\left(\zeta + \frac{1}{4\zeta}\right) \text{ Hz}$$

INPUT	FINAL ERROR, R_ϵ
ΔR	0
ΔVELOCITY	0
ΔACCELERATION	$\Delta ACCEL/\omega_n^2$

FIGURE 3-10. Second-Order, Type II, Range-Tracker General Equation Summary.

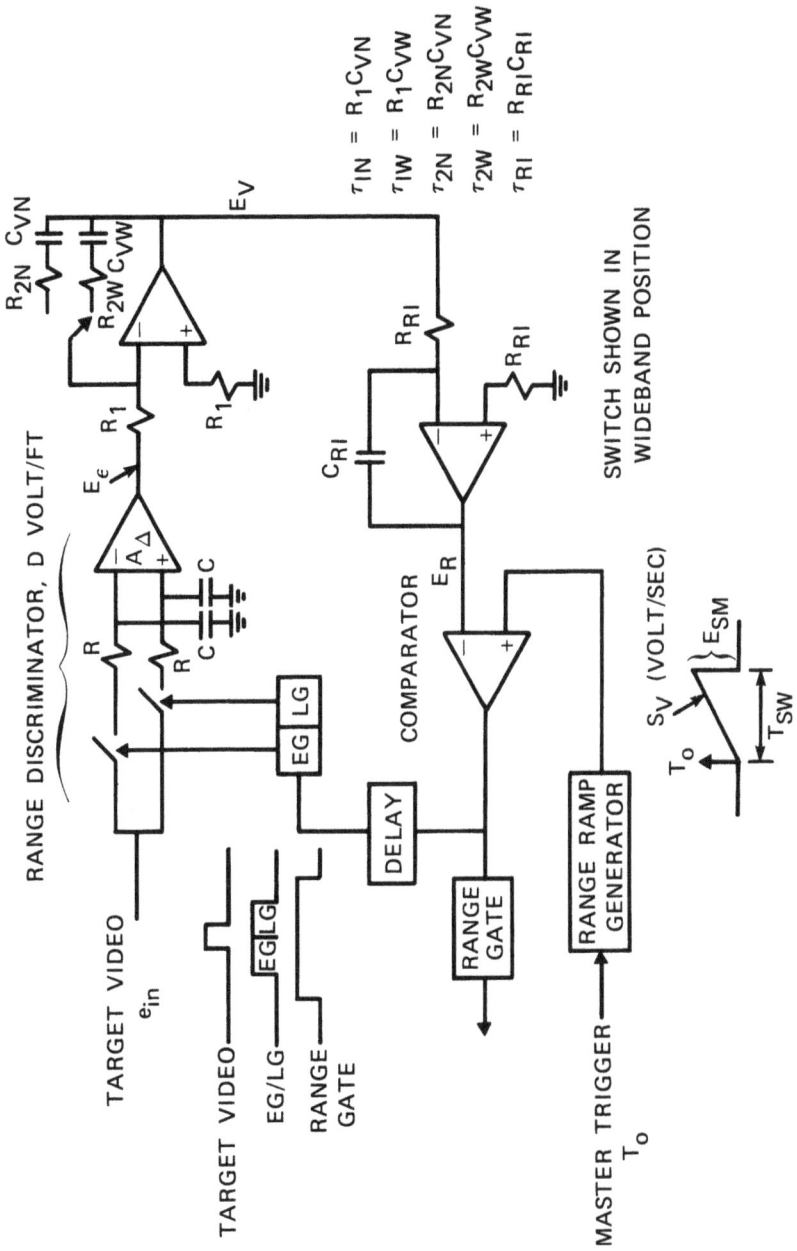

RANGE DISCRIMINATOR, D VOLT/FT

$$\tau_{IN} = R_1 C_{VN}$$
$$\tau_{IW} = R_1 C_{VW}$$
$$\tau_{2N} = R_{2N} C_{VN}$$
$$\tau_{2W} = R_{2W} C_{VW}$$
$$\tau_{RI} = R_{RI} C_{RI}$$

SWITCH SHOWN IN
WIDEBAND POSITION

C. Calculate the range integrator time constant, τ_{RI}. The range integrator output voltage may be given as*

$$\frac{d\,E_R}{dt} = \frac{E_v}{\tau_{RI}} \qquad (3\text{-}25)$$

Since E_v represents the target range-tracker closing velocity, the range voltage will be a linear ramp (remember a Type II loop has zero velocity error, and the range voltage, E_R, must keep up with the target velocity). Referring to Figure 3-5, the velocity of R_o must be the same as the velocity of R_{in} (Vel_{in}); thus,

$$\frac{\dfrac{d\,R_o}{dt}}{S_R} = \frac{d\,E_R}{dt} = \frac{E_v}{\tau_{RI}} \qquad (3\text{-}26)$$

or

$$\frac{Velocity}{S_R} = \frac{E_v}{\tau_{RI}} \qquad (3\text{-}27)$$

Knowing the maximum velocity (this must be given as a system specification), we are free to choose the maximum value for E_v. Thus τ_{RI} may now be found.

$$\tau_{RI} = R_{RI}\,C_{RI} = \frac{E_v(\text{max})\,S_R}{Vel(\text{max})} \qquad (3\text{-}28)$$

D. The wideband velocity integrator, τ_{1W} and τ_{2W}, may be found knowing the maximum velocity step (ΔVel), the desired system damping factor (ζ), S_R, the discriminator scale factor (D), and the maximum range error (R_v). Using Figure 3-8, find the maximum value (for the desired system ζ) for

---

*The minus sign for the integrator will be omitted as discussed earlier.

$[R_\varepsilon(f)]/[\Delta Vel/\omega_n]$. Let X_{vel} equal this value. The wideband system ω_{nw} may now be found as

$$\omega_{nw} = \frac{X_{vel}\,\Delta Vel}{R_\varepsilon(max)} \qquad (3\text{-}29)$$

From Figure 3-10,

$$\omega_{nw}^{\;2} = \frac{DS_R}{\iota_{RI}\,\iota_{1w}} \qquad (3\text{-}30)$$

or

$$\iota_{1w} = R_1\,C_{vw} = \frac{DS_R}{\iota_{RI}\,\omega_{nw}^2} \qquad (3\text{-}31)$$

also,

$$\zeta = \frac{\iota_{2w}\,\omega_{nw}}{2} \qquad (3\text{-}32)$$

or

$$\iota_{2w} = R_{2w}\,C_{vw} = \frac{2\zeta}{\omega_{nw}} \qquad (3\text{-}33)$$

The loop closure time, t_c, may be found knowing ω_n , ζ, and using Figure 3-8 as follows: find $\omega_n t$ where the relative range error goes to zero and divide by ω_{nw}.

E. Knowing ω_{nw} and ζ, the loop noise bandwidth may be found:

$$BW_{nw} = \frac{\omega_{nw}}{2}\left(\zeta + \frac{1}{4\zeta}\right) \text{Hz} \qquad (3\text{-}34)$$

F. The loop bandwidth can be decreased after the target has been acquired in the wideband mode. Decreasing the loop bandwidth has the advantage of making range gate stealing (a common countermeasure

191

technique) more difficult. The loop should just be fast enough to follow any acceleration changes. Referring to Figure 3-9, find the maximum value for $[R_\varepsilon(t)]/[\Delta Accel/\omega_n{}^2]$ for the system ζ. Call this value:

$$X_{Accel} = \frac{R_\varepsilon(max)}{\Delta Accel/\omega_{nw}{}^2} \tag{3-35}$$

The narrowband loop natural frequency may now be found:

$$\omega_{nN} = \sqrt{\frac{X_{Accel} \, \Delta Accel}{R_\varepsilon(max)}} \tag{3-36}$$

Now, from Figure 3-10

$$\iota_{1N} = R_1 \, C_{VN} = \frac{D \, S_R}{\iota_{RI} \, \omega_{nN}{}^2} \tag{3-37}$$

The value for R_1 is determined from the wideband case and a unique value of C_{VN} can be found.

$$C_{VN} = \frac{D \, S_R}{\iota_{RI} \, \omega_{nN}{}^2 \, R_1} \tag{3-38}$$

The system damping ratio, ζ, should be independent of ω_n, thus a unique value of R_{2N} may be determined.

$$\zeta = \frac{\iota_{2N}}{2} \, \omega_{nN} \tag{3-39}$$

Solving for ι_{2N}

$$\iota_{2N} = R_{2N} \, C_{VN} = \frac{2\zeta}{\omega_{nN}} \tag{3-40}$$

thus,

$$R_{2N} = \frac{2\zeta}{\omega_{nN} \, C_{VN}} \tag{3-41}$$

192

The final range error is

$$R_\varepsilon(\text{final}) = \frac{\Delta\text{Accel}}{\omega_{nN}^2}$$ (3-42)

G. Knowing ω_{nN} and ζ, the loop noise bandwidth may be found

$$BW_{NN} = \frac{\omega_{nN}}{2}\left(\zeta + \frac{1}{4\zeta}\right) \text{Hz}$$ (3-43)

H. The range discriminator is discussed in Appendix 3B, and the equations presented are excellent approximations; however, the discriminator differencing amplifier gain, A_Δ, may have to be adjusted to give the desired scale factor. Practice has shown that a value for D of 50 mV/ft gives reasonable values for the range and velocity integrator component values. Thus, practical designs assume a value for D, and calculating the range discriminator differencing amplifier gain, A_Δ, is necessary to give D. The value for A_Δ may be given as (Appendix 3B)

$$A_\Delta = \frac{D\,t_{EG}\,(\text{nsec})}{4.1\,e_{in}}$$ (3-44)

Figures 3-12 through 3-16 summarize the design procedure as an easy reference for the design to follow.

193

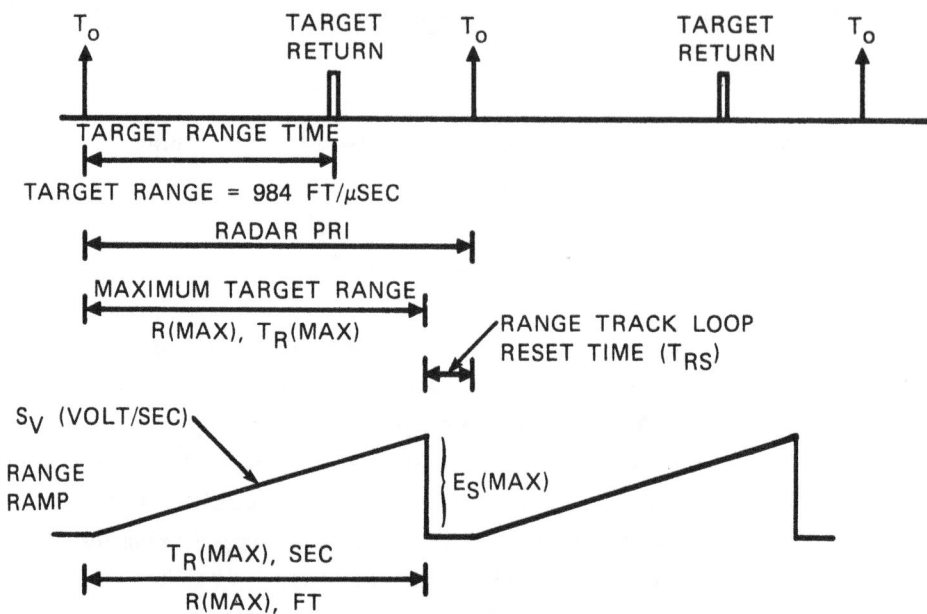

$$T_R(MAX) = 2.03 \times 10^{-9} \; R_{MAX}(FT) \quad SEC$$

$$S_V = \frac{E_S(MAX)}{T_R(MAX)} \quad VOLT/SEC$$

$$S_R = \frac{R_{MAX}(FT)}{E_S(MAX)} \quad FT/VOLT$$

$$S_R = \frac{R_{MAX}(FT)}{S_V \; T_R(MAX)} \quad FT/VOLT$$

THE RANGE TRACK LOOP RESET TIME ALLOWS RANGE RAMP RESET AND INITIALIZATION, AND, DEPENDING ON DESIGN, IS IN THE ORDER OF 10 μSEC.

$$PRI = \frac{1}{PRF} = T_R(MAX) + RESET \; TIME \; (T_{RS})$$

FIGURE 3-12. Range Tracking Loop Scale Factor Design Equations.

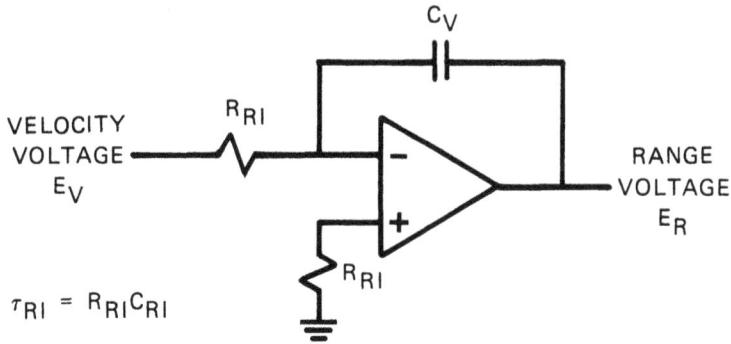

$\tau_{RI} = R_{RI}C_{RI}$

$E_V(MAX)$ = VELOCITY VOLTAGE FOR MAXIMUM VELOCITY (VEL(MAX))

VEL(MAX) = MAXIMUM VELOCITY

S_R = RANGE INTEGRATOR SCALE FACTOR, FT/VOLT

$$\tau_{RI} = R_{RI}C_{RI} = \frac{E_V(MAX)\, S_R}{VEL(MAX)}$$

FIGURE 3-13. Range Integrator Design Equations.

Design Example:

A range-tracking loop will be designed to meet the following specifications:

Maximum range, $R(\text{max})$	180×10^3 ft
Transmitted pulse width	200 nsec
Maximum closing velocity, $Vel(\text{max})$	1250 ft/sec
Maximum velocity step, ΔVel	400 ft/sec
Maximum acceleration, $\Delta Accel(\text{max})$	5.6 g (180 ft/sec^2)
Early/Late gate width, t_{EG}	250 nsec
Maximum range error due to ΔVel, R_ε	10 feet
Maximum range error due to $\Delta Accel$, R_ε	5 feet
System damping factor, ζ	0.7
Normalized video input, e_{in}	2 volts

195

$$\tau_{1W} = R_1 C_{VW}$$
$$\tau_{2W} = R_{2W} C_{VW}$$

FINAL RANGE ERROR = 0

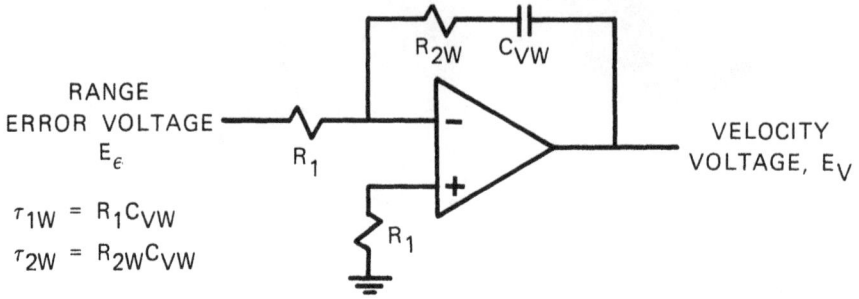

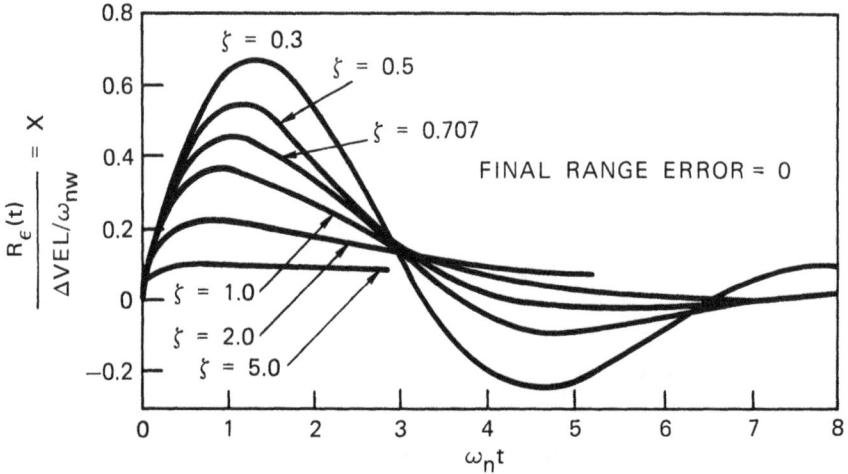

FIND THE MAXIMUM VALUE OF X, FOR THE SYSTEM ζ, FROM THE ABOVE GRAPH.

$$\omega_{nW} = \frac{X \, \Delta VEL}{R_\varepsilon(MAX)}$$

$$R_1 C_{VW} = \frac{D \, S_R}{\tau_{RI} \, \omega_{nW}^2} \qquad R_{2W} C_{VW} = \frac{2\zeta}{\omega_{nW}}$$

$$BW_{nW} = \frac{\omega_{nW}}{2} \left(\zeta + \frac{1}{4\,\zeta} \right) \text{ Hz}$$

THE LOOP CLOSURE TIME MAY BE FOUND, KNOWING ζ AND ω_{nN}, FROM THE ABOVE GRAPH.

FIGURE 3-14. Velocity Integrator Design Equations (Wideband).

196

RANGE ERROR
VOLTAGE, E_ϵ

R_1

R_{2N} C_{VN}

VELOCITY
VOLTAGE, E_V

$\tau_{1N} = R_1 C_{VN}$
$\tau_{2N} = R_{2N} C_{VN}$

R_1

$\dfrac{R_\epsilon(t)}{\Delta ACCEL/\omega_{nN}^2} = X$

$\zeta = 0.3$
$\zeta = 0.5$
$\zeta = 0.707$
$\zeta = 1.0$
$\zeta = 2.0$

FINAL
RANGE ERROR $= \dfrac{\Delta ACCEL}{(\omega_{nN})^2}$

$\zeta = 5.0$

$\omega_n t$

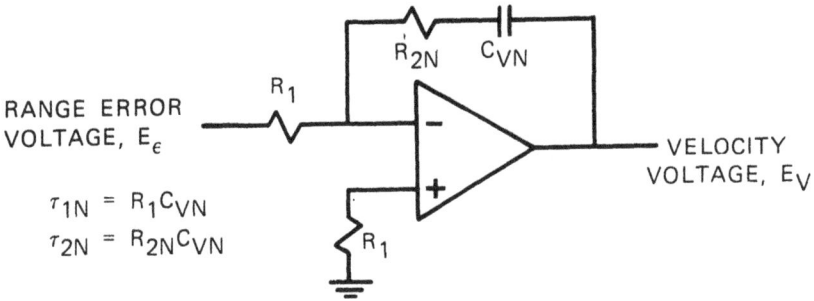

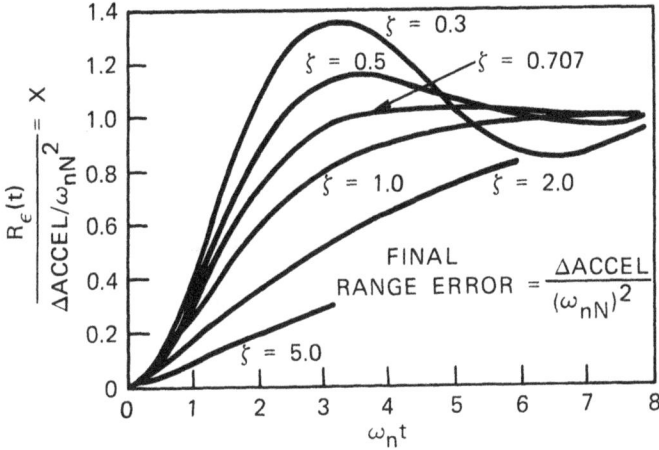

FIND THE MAXIMUM VALUE OF X, FOR THE SYSTEM ζ,
FROM THE ABOVE GRAPH.

$$\omega_{nN} = \sqrt{\frac{X \Delta ACCEL}{R_\epsilon(MAX)}}$$

$$R_1 C_{VN} = \frac{D S_R}{\iota_{RI}\, \omega_{nN}^2} \qquad R_{2N} = \frac{2\zeta}{\omega_{nN} C_{VN}}$$

$$BW_{nW} = \frac{\omega_{nW}}{2}\left(\zeta + \frac{1}{4\zeta}\right) \; Hz$$

THE LOOP CLOSURE TIME MAY BE FOUND, KNOWING ζ AND ω_{nw},
FROM THE ABOVE GRAPH.

FIGURE 3-15. Velocity Integrator Design Equations (Narrowband).

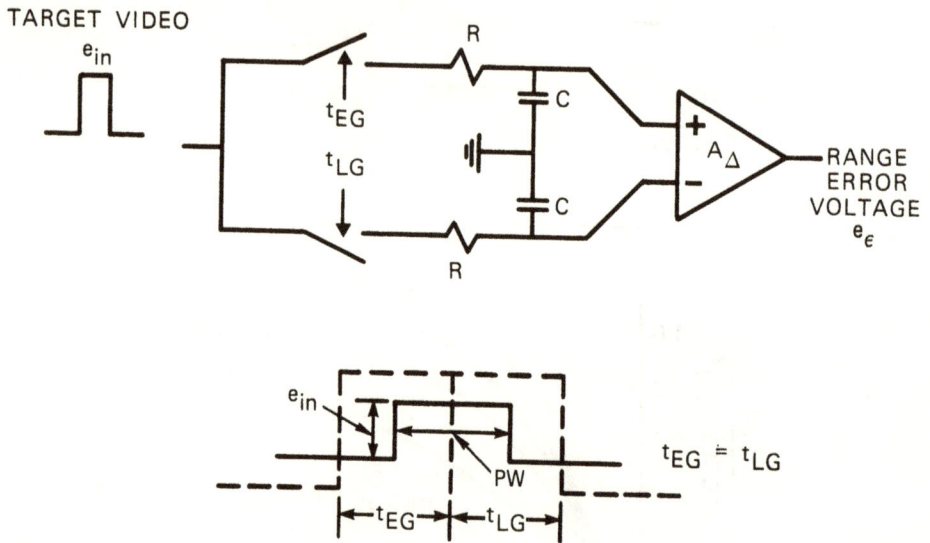

$$RC > 10\,t_{EG} \qquad\qquad RC < 0.02\,t_c\,t_{EG}\,PRF$$

t_c = LOOP CLOSURE TIME

PRF = RADAR PULSE REPETITION FREQUENCY

DISCRIMINATOR SCALE FACTOR, SE_ϵ

$$SE_\epsilon = \frac{2\,A_\Delta\,e_{in}}{t_{EG}\,(nsec)} \quad \frac{VOLT}{NSEC}$$

SYSTEM RANGE DISCRIMINATOR SCALE FACTOR, D

$$D = SE_\epsilon\left(\frac{volt}{nsec}\right)\,2.03\,\frac{nsec}{ft} = \frac{4.06\,A_\Delta\,e_{in}}{t_{EG}\,(nsec)}\,\frac{VOLT}{FT}$$

FIGURE 3-16. Range Discriminator Design Equations.

Determine $T_R(max)$, S_v, S_R, and PRF (Figure 3-12)

$$T_R(max) = 2.03 \times 10^{-9}\ [R(max)] = 2.03 \times 10^{-9}\ (180 \times 10^3) \tag{3-45}$$

$$T_R(max) = 365.8\ \mu sec \tag{3-46}$$

Let the maximum range sweep voltage, $S_E(max) = 15$ volts

$$S_v = \frac{E_s(max)}{T_R(max)} = \frac{15}{365.8 \times 10^{-6}} \tag{3-47}$$

$$S_v = 41 \times 10^3\ V/sec \tag{3-48}$$

$$S_R = \frac{R(max)}{E_s(max)} = \frac{180 \times 10^3}{15} \tag{3-49}$$

$$S_R = 12{,}000\ ft/V \tag{3-50}$$

$$PRI = T_R(max) + T_{RS} \tag{3-51}$$

Let the system reset time $= 34.2\ \mu sec$ (a more than reasonable value).*

$$PRI = 365.8\ \mu sec + 34.2\ \mu sec = 400\ \mu sec \tag{3-52}$$

$$PRF = \frac{1}{400\ \mu sec} = 2.5\ kHz \tag{3-53}$$

B. Calculate R_{RI} and C_{RI} (Figure 3-13). Let $E_v(max) = 10$ volts for a $Vel(max) = 1250$ ft/sec

$$R_{RI}\,C_{RI} = \frac{E_v(max)\,S_R}{Vel(max)} = \frac{10\,(12{,}000)}{1250} \tag{3-54}$$

$$R_{RI}\,C_{RI} = 96 \tag{3-55}$$

---

*Seldom is the PRF determined by the system reset time; however, sufficient system reset time must be accounted for in the design.

Equation (3-55) yields unrealistic values, since, for a C_{RI} of 1 μF

$$R_{RI} = \frac{96}{1 \times 10^{-6}} = 96\ M\Omega \tag{3-56}$$

The problem may be corrected by attenuating E_V as shown in Figure 3-17.

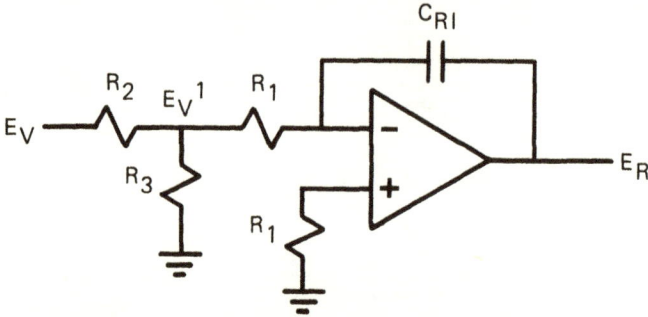

FIGURE 3-17. Range Integrator Variation.

If $R_3 \ll R_1$

$$E_v' = \frac{R_3\ E_v}{R_2 + R_3} \tag{3-57}$$

or

$$\frac{E_v'}{E_v} = \frac{R_3}{R_2 + R_3} = Atten. \tag{3-58}$$

Let $C_{RI} = 2.2\ \mu F$ and $R_1 = 1M\Omega$

$$R_1\ C_{RI} = \frac{E_v'(max)\ S_R}{Vel(max)} = 2.2 \tag{3-59}$$

$$E_v' = \frac{R_1\ C_{RI}\ Vel(max)}{S_R} \tag{3-60}$$

$$E_v' = \frac{2.2\,(1250)}{12,000} = 0.229 \qquad (3\text{-}61)$$

$$\frac{E_v'}{E_v} = \frac{R_3}{R_2 + R_3} \qquad (3\text{-}62)$$

Let $R_3 = 1\,k\Omega$ (this is much less than R_1)

$$R_2 = \frac{R_3\,(1 - E_v'/E_v)}{E_v'/E_v} \qquad (3\text{-}63)$$

$$R_2 = \frac{1 \times 10^3\,(1 - 0.229/10)}{0.229/10} = 42.6\,k\Omega \qquad (3\text{-}64)$$

To verify this result,

$$\iota_{RI} = \frac{R_1 C_{RI}}{\dfrac{R_3}{R_3 + R_2}} \qquad (3\text{-}65)$$

$$\iota_{RI} = \frac{(1 \times 10^6)(2.2 \times 10^{-6})}{\dfrac{1 \times 10^3}{(42.6 \times 10^3) + (1 \times 10^3)}} = 96 \qquad (3\text{-}66)$$

C. Calculate wideband R_1, R_{2w}, C_{vw} and BW_{nw} (Figure 3-14). The maximum value of X, for $\zeta = 0.7$ is (Figure 3-14)

$$X \simeq 0.45 \qquad (3\text{-}67)$$

$$\omega_{nw} = \frac{X \Delta \mathrm{Vel}}{R_\varepsilon(\max)} = \frac{(0.46)(400)}{10} \qquad (3\text{-}68)$$

201

$$\omega_{nw} = 18.4 \text{ rad/sec} \tag{3-69}$$

$$R_1 C_{vw} = \frac{D S_R}{\tau_{RI} \omega_n^2} \tag{3-70}$$

Let $D = 50 \text{ mV/ft}$

$$R_1 C_{vw} = \frac{(0.05)(12,000)}{97(18.4)^2} = 18.27 \times 10^{-3} \tag{3-71}$$

Let $C_{vw} = 0.25 \text{ } \mu\text{F}$

$$R_1 = \frac{18.27 \times 10^{-3}}{0.25 \times 10^{-6}} = 73 \text{ k}\Omega \tag{3-72}$$

$$R_{2w} = \frac{2\zeta}{\omega_{nw} C_{vw}} \tag{3-73}$$

$$R_{2w} = \frac{2(0.707)}{(18.4)(0.25 \times 10^{-6})} = 304 \text{ k}\Omega \tag{3-74}$$

$$BW_{nw} = \frac{\omega_{nw}}{2} \left(\zeta + \frac{1}{4\zeta} \right) \text{ Hz} \tag{3-75}$$

$$BW_{nw} = \frac{18.4}{2} \left(0.7 + \frac{1}{4(0.7)} \right) = 9.7 \text{ Hz} \tag{3-76}$$

The loop closure time for a step velocity input can be found for the loop ζ by noting the value of $\omega_n t$ for the normalized range error, X, to cross zero. From Figure 3-14 it can be seen that for $\zeta = 0.7$,

$$W_n t \simeq 3.8 \tag{3-77}$$

the wideband loop closure time is

$$t_{cw} = \frac{3.8}{\omega_{nw}} = \frac{3.8}{18.4} \tag{3-78}$$

$$t_{cw} = 0.21 \text{ sec} \tag{3-79}$$

D. Calculate narrowband R_{2N}, C_{vN} and BW_{nN} (Figure 3-15). The maximum value of X, for $\zeta = 0.7$, is (Figure 3-15)

$$X \simeq 1.05 \tag{3-80}$$

$$\omega_{nN} = \sqrt{\frac{X \Delta \text{Accel}}{R_\varepsilon(\text{max})}} = \sqrt{\frac{(1.05)(180)}{5}} \tag{3-81}$$

$$\omega_{nN} = 6.15 \text{ rad/sec} \tag{3-82}$$

$$C_{vn} = \frac{D S_R}{R_1 t_{RI} (\omega_{nN})^2} = \frac{(50 \times 10^{-3})(12 \times 10^3)}{(73 \times 10^3)(97)(6.15)^2} \tag{3-83}$$

$$C_{vn} = 2.24 \text{ μF} \tag{3-84}$$

$$R_{2N} = \frac{2\zeta}{\omega_{nN} C_{vN}} = \frac{2(0.70)}{(6.15)(2.24 \times 10^{-6})} \tag{3-85}$$

$$R_{2N} = 102 \text{ k}\Omega \tag{3-86}$$

$$BW_{nN} = \frac{\omega_{nN}}{2}\left(\zeta + \frac{1}{4\zeta}\right) = \frac{6.15}{2}\left(0.7 + \frac{1}{4(0.7)}\right) \tag{3-87}$$

$$BW_{nN} = 3.2 \text{ Hz} \tag{3-88}$$

203

The final range error due to acceleration is

$$R_\varepsilon \text{(final)} = \frac{\Delta \text{Accel}}{(\omega_{nN})^2} = \frac{180}{(6.15)^2} \qquad (3\text{-}89)$$

$$R_\varepsilon \text{(final)} = 4.7 \text{ ft} \qquad (3\text{-}90)$$

The loop closure time may be found, for the loop ζ, by noting the value of $\omega_n t$ for the normalized range error, X, to cross zero in Figure 3-15. For $\zeta = 0.7$

$$\omega_n t \simeq 3.2 \qquad (3\text{-}91)$$

$$t_{CN} = \frac{3.2}{\omega_{nN}} = \frac{3.2}{6.15} \qquad (3\text{-}92)$$

$$t_{CN} = 0.52 \text{ sec} \qquad (3\text{-}93)$$

E. Calculate range discriminator R, C, and A_Δ (Figure 3-16)

$$RC > 10\, t_{EG} \quad RC > 10\,(250 \times 10^{-9}) \qquad (3\text{-}94)$$

$$RC > 2.5 \times 10^{-6} \qquad (3\text{-}95)$$

$$RC < 0.02\, t_c\, t_{EG}\, \text{PRF} \qquad (3\text{-}96)$$

The minimum value for the loop closure time, t_c, will be used ($t_{CW} = 0.21$ second)

$$RC < 0.02\,(0.21)(250 \times 10^{-9})(2.5 \times 10^3) \qquad (3\text{-}97)$$

$$RC < 2.6 \times 10^{-6} \qquad (3\text{-}98)$$

204

Let $C = 0.001 \, \mu F$ $R = 2.5 \, k\Omega$

$$RC = 2.5 \times 10^{-6} \qquad \text{(3-99)}$$

If RC is less than 10 t_{EG}, the discriminator scale factor will be low (which is easily corrected by adjusting A_Δ for the proper value). If RC is greater than 0.02 t_c t_{EG} PRF, range track loop dynamics may be affected

The range discriminator gain may now be found (Figure 3-16).

$$A_\Delta = \frac{D \, t_{EG} \, (nsec)}{4.06 \, e_N} \qquad \text{(3-100)}$$

$$A_\Delta = \frac{(0.05)(250)}{(4.06)(2)} = 1.54 \qquad \text{(3-101)}$$

F. Design the range ramp generator.

There are many methods to generate a linear voltage ramp. One straightforward method is illustrated in Figure 3-18. The switch is closed during the reset time, T_{RS} (Figure 3-12), and the range ramp voltage is zero. At time, T_o, the switch opens and the output linearly ramps up according to the following equation:

$$\frac{d \, E_{out}}{dt} = -\left(\frac{-V}{RC} \right) = \frac{V}{RC} \, V/sec \qquad \text{(3-102)}$$

The output slope is (Figure 3-12)

$$S_V = \frac{V}{RC} \, V/sec \qquad \text{(3-103)}$$

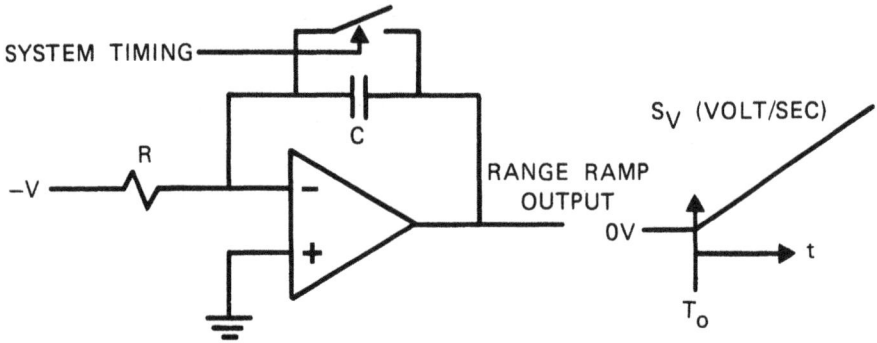

FIGURE 3-18. Range Ramp Generator.

The system S_v has been previously determined,

$$S_V = 41 \times 10^3 \text{ V/sec} \tag{3-104}$$

Let $V = -10$ volts and $C = 0.1 \ \mu F$

$$R = \frac{V}{S_V C} = \frac{10}{(41 \times 10^3)(0.1 \times 10^{-6})} \tag{3-105}$$

$$R = 2.44 \text{ k}\Omega \tag{3-106}$$

It is a simple task to adjust R to give the exact S_V needed.

Figures 3-19 and 3-20 illustrate the range tracking loop. Figure 3-19 is straightforward, but a few words are warranted about the comparator (Figure 3-20). The range integrator scale factor is

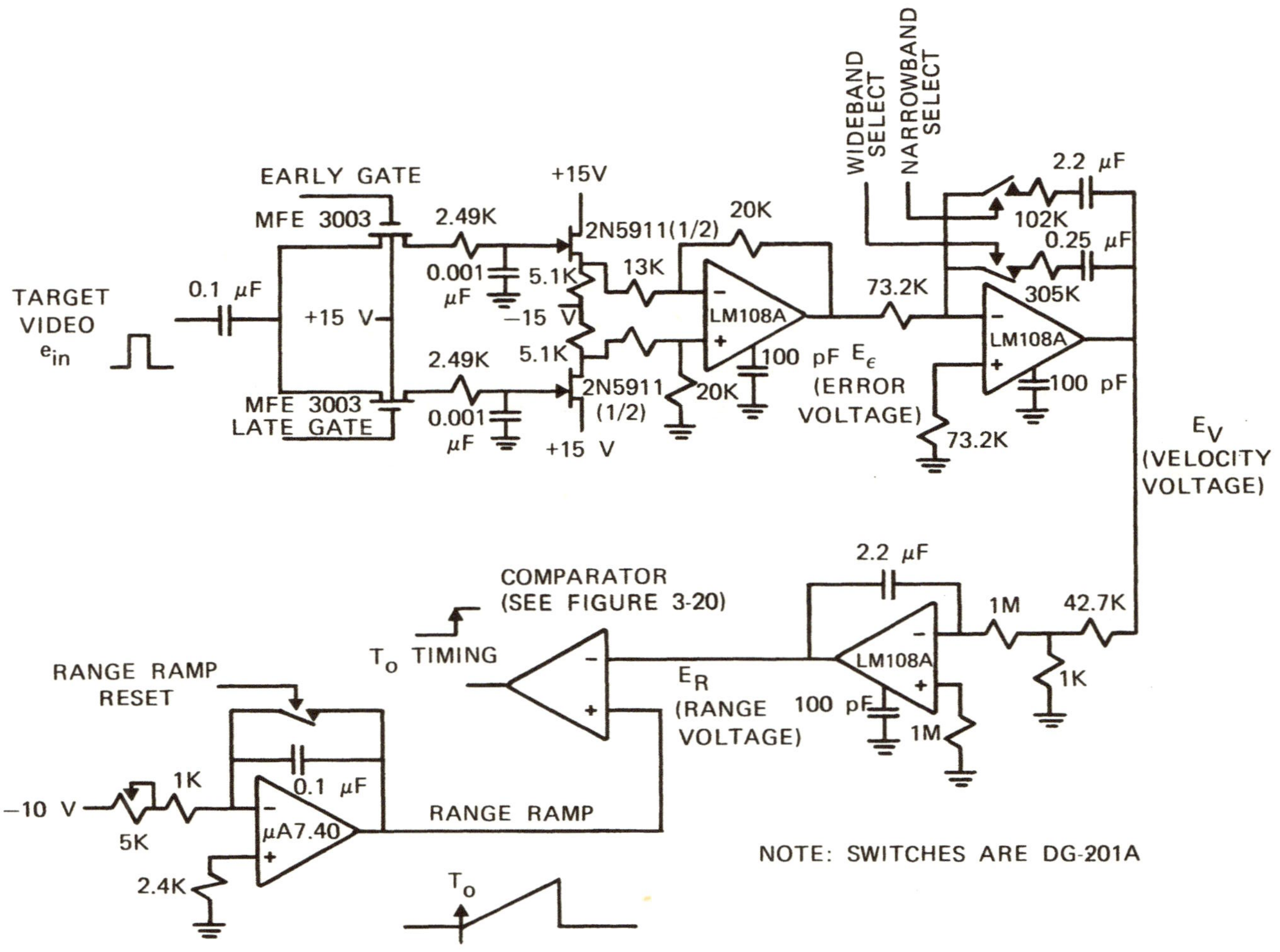

FIGURE 3-19. Second-Order, Type II, Range Track Loop.

207

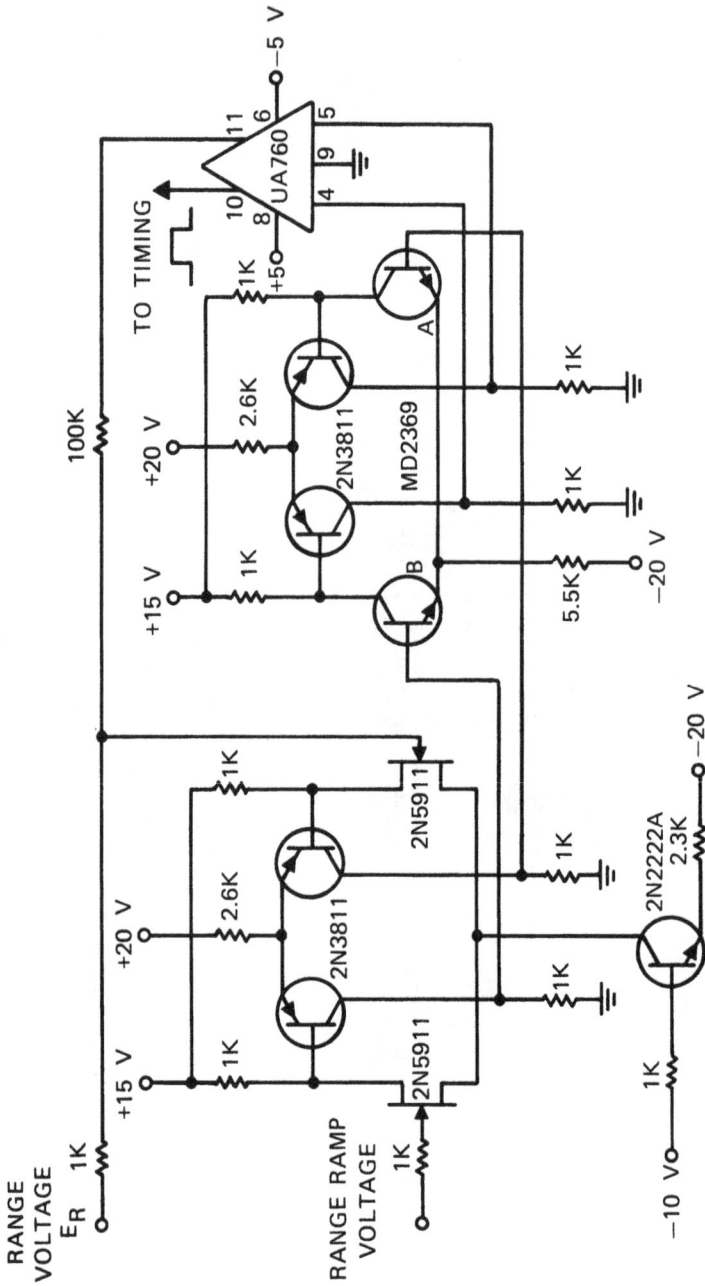

FIGURE 3-20. Comparator.

$$S_R = 12,000 \text{ ft/V} \qquad (3\text{-}107)$$

or put in a different perspective

$$S_R = 12 \text{ ft/mV} \qquad (3\text{-}108)$$

To avoid any range timing jitter, the comparator must be quite accurate (1 mV of comparator uncertainty represents 12 feet of range uncertainty, or range jitter). The comparator shown in Figure 3-19 reduces range jitter to less than 5 feet.

The loop natural frequency, ω_n, and damping, ζ, are a function of the discriminator scale factor, D, which in turn is a function of the target video return, e_{in}. There is generally some variation in e_{in} due to target modulation (see Chapter 1 for AGC limitations and Reference 3 if the range track loop is driven by a logarithmic amplifier). Figure 3-21 shows the effect on ω_n and ζ due to variations in D. Table 3-2 summarizes the design.

Verifying a range tracking loop design requires test equipment seldom available except in specialized laboratories. A simple method, however, exists to measure a range track loop's natural frequency and damping. It is a fairly simple matter to simulate a target return pulse at some arbitrary time (range) from a simulated transmit pulse (see Figure 3-12). Stepping the simulated return ± 50 nsec simulates a range change of ± 24.6 feet (range = 0.49 ft/nsec). Measuring the range error voltage, E_{ε} (Figure 3-19), a curve similar to Figure 3-7 is obtained. There is a unique relationship between the first zero crossing and the percent of overshoot (Figure 3-22), and the loop's natural frequency, ω_n, and damping, ζ. Figure 3-23 is a plot of ζ and $W_n t_0$ for a measured percent of overshoot. The usefulness of Figure 3-23 is shown by the following:

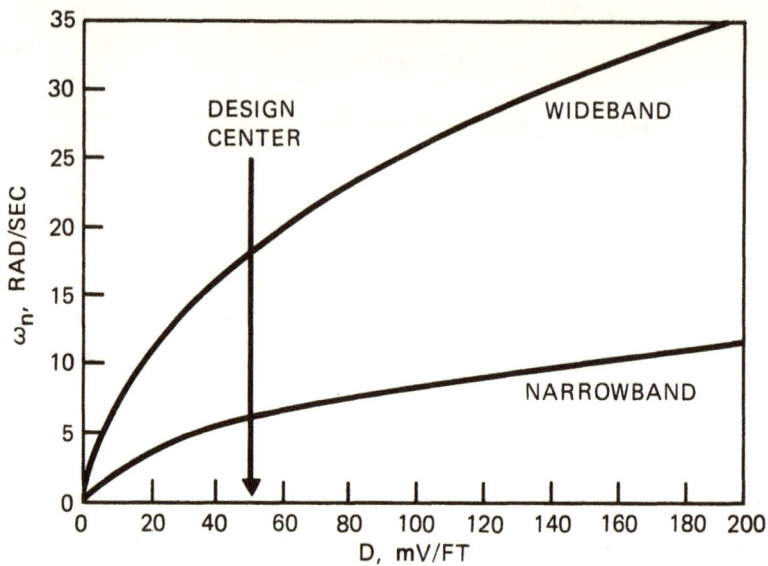

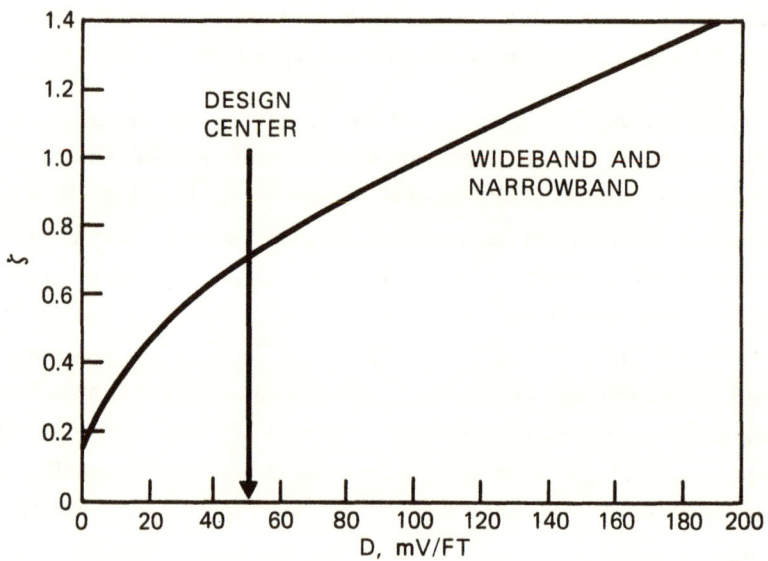

FIGURE 3-21 Effect of Varying D on Natural
Frequency, ω_n, and Damping, ζ.

Table 3-2. Design Summary.

	Wide bandwidth	Narrow bandwidth
ζ	0.707	0.707
ω_n, rad/sec	18.4	6.15
BW_n, Hz	9.7	3.2
Loop closure time, sec	0.21	0.52

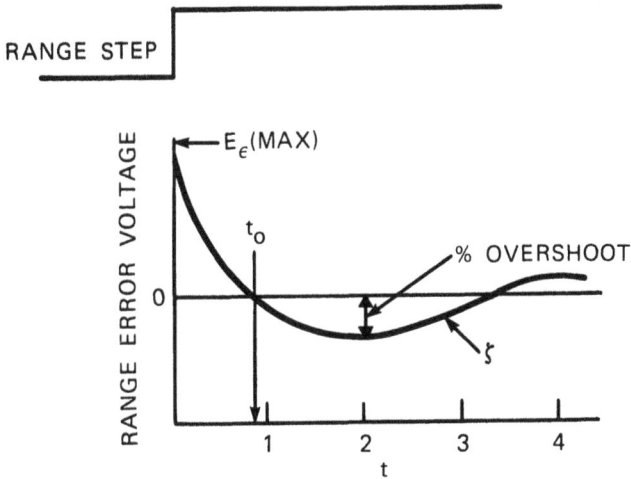

FIGURE 3-22. Range Error Voltage, $R_\varepsilon(t)$, Due to a Step Change in Range (See Figure 3-7).

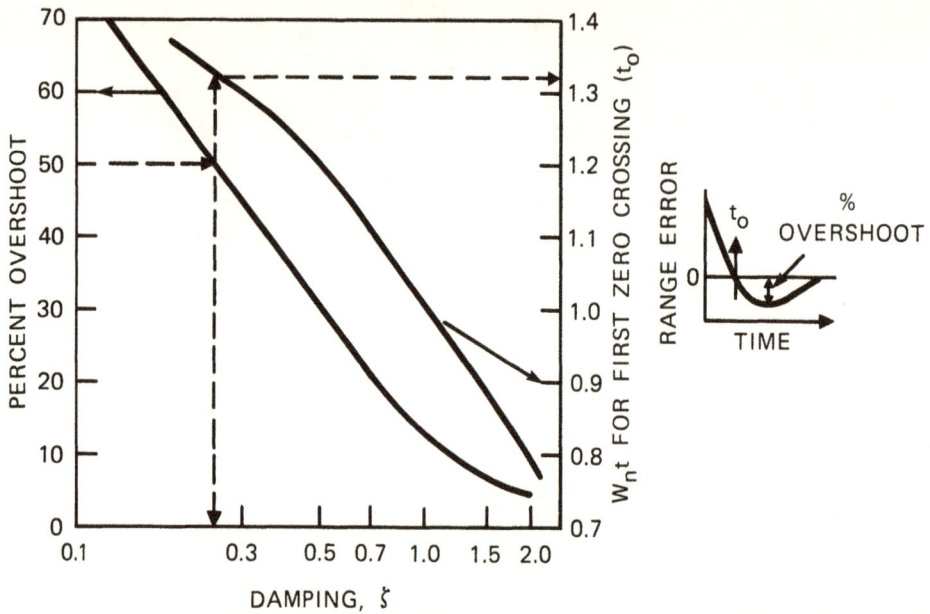

FIGURE 3-23. Overshoot and $\omega_n t_0$ Versus Damping (ζ).

Given % overshoot = 50%

$$t_0 = 0.3 \text{ second}$$

$$\zeta = 0.25$$

$$\omega_n t_0 = 1.32$$

$$\omega_n = \frac{1.32}{0.3} = 4.4 \text{ rad/sec}$$

The measured results of the loop are within 5% of the designed values.

Some designs use a Type I range tracking loop (finite velocity error), and the basic design procedure will now be covered.

Second Order, Type I, Ranging Tracking Loop

The loop filter illustrated in Figure 3-24 will now be discussed. The loop filter transfer function may be given as (A_V is included for amplitude scaling or buffering as necessary).

$$F(S) = A_V \frac{(1 + \tau_2 S)}{(1 + \tau_1 S)} \qquad (3\text{-}109)$$

Substituting Equation (3-109) into Equation (3-3), the loop gain is

$$LG(S) = \frac{DS_R A_V}{\tau_{RI} S} \left[\frac{1 + \tau_2 S}{1 + \tau_1 S} \right] \qquad (3\text{-}110)$$

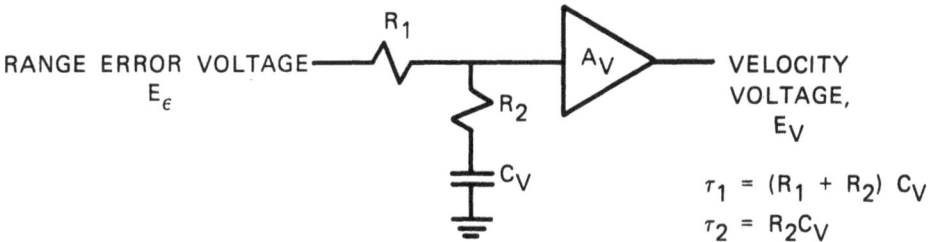

RANGE ERROR VOLTAGE E_ϵ — R_1 — A_V — VELOCITY VOLTAGE, E_V

R_2 C_V

$$\tau_1 = (R_1 + R_2) C_V$$
$$\tau_2 = R_2 C_V$$

FIGURE 3-24. Type I Loop Filter $F(S) = A_V [(1 + \tau_2 S)/(1 + \tau_1 S)]$.

The denominator of Equation (3-110) has a single pole at the origin, thus, this is a Type I loop (finite velocity error). The loop transfer function (Equation (3-5)) may be given as

$$H(S) = \frac{\left(\dfrac{DS_R A_V \tau_2}{\tau_{RI} \tau_1} \right) S + \dfrac{DS_R A_V}{\tau_{RI} \tau_1}}{S^2 + \left(\dfrac{1}{\tau} + \dfrac{DS_R A_V T_2}{\tau_1 \tau_{RI}} \right) S + \dfrac{DS_R A_V}{\tau_{RI} \tau_1}} \qquad (3\text{-}111)$$

Defining

$$\frac{DS_R A_V}{\tau_{RI}\tau_1} = \omega_n^2 \qquad (3\text{-}112)$$

or

$$\omega_n = \sqrt{\frac{DS_R A_V}{\tau_{RI}\tau_1}} \qquad (3\text{-}113)$$

and

$$\frac{1}{\tau_1} + \frac{DS_R A_V \tau_2}{\tau_1 \tau_{RI}} = 2\zeta\omega_n \qquad (3\text{-}114)$$

Solving for ζ,

$$\zeta = \frac{1}{2\omega_n\tau_1}\left(1 + \frac{DS_R A_V \tau_2}{\tau_{RI}}\right) \qquad (3\text{-}115)$$

Substituting Equations (3-115) and (3-112) into Equation (3-111), H(S) may be given as

$$H(S) = \frac{\omega_n^2 + \left(2\zeta\omega_n - \dfrac{\omega_n^2 \tau_{RI}}{DS_R A_V}\right)S}{S^2 + 2\zeta\omega_n + \omega_n^2} \qquad (3\text{-}116)$$

The normalized range error, $R_\varepsilon(S)$ (Equation (3-8)), may now be found:

$$\frac{R_\varepsilon(S)}{R_{in}(S)} = \frac{S^2 + \omega_n^2\left(\dfrac{\tau_{RI}}{DS_R A_V}\right)S}{S^2 + 2\zeta\omega_n S + \omega_n^2} \qquad (3\text{-}117)$$

The loop dynamics for step changes in range, velocity, and acceleration (Equation (3-10)) are quite complex [1], however, in summary:

A. Step change in range (ΔR). Provided

$$\frac{D S_R A_V}{\iota_{RI}} \gg \frac{\omega_n}{\zeta} \tag{3-118}$$

the curves generated for the Type II loop (Figures 3-7 and 3-23) are valid for the Type I loop. The final range error is

$$R_\varepsilon \text{ (final)} = 0 \tag{3-119}$$

B. Step change in velocity (ΔVel). The final range error is

$$R_\varepsilon \text{ (final)} = \frac{\Delta Vel}{\dfrac{D S_R A_V}{\iota_{RI}}} \tag{3-120}$$

and provided $R_\varepsilon(final)$ is small (which can be ensured by making A_V large), the curves generated for the Type II loop (Figure 3-8) are valid.

C. Step change in acceleration ($\Delta Accel$). If $(D\ S_R A_V)/\iota_{RI}$ is large, the curves generally follow those of the Type II loop (Figure 3-9); however, the range error increases with time which will eventually cause loss of lock.

$$R_\varepsilon(t) \simeq \frac{(\Delta Accel)(t)}{\dfrac{D S_R A_V}{\iota_{RI}}} \tag{3-121}$$

The design procedure is similar to that of the Type II loop.

Type I Tracking Loop Design Equations

A. Knowing R_{max} (ft) determines the range ramp sweep width (T_{SW}) and scale factors, S_V and S_R. This procedure is exactly the same as for the Type II loop (see Figure 3-12).

215

B. Calculate the range integrator time constant, t_{RI}. This differs considerably from Type II design as there is a finite range error as a function of velocity. The range integrator voltage must meet the same criteria as the Type II loop (Equation (3-26)),

$$\frac{d\,E_R}{d\,t} = \frac{E_V(max)}{t_{RI}} = \frac{Vel(max)}{S_R} \tag{3-122}$$

(it will be assumed that A_V is unity, as this is most often the case for practical Type I systems).

$$E_\varepsilon(max) = R_\varepsilon(Vel)\,D = E_V(max) \tag{3-123}$$

where

$$R_\varepsilon(Vel) = \text{Velocity range error}$$

Solving Equations (3-121) and (3-122) for t_{RI}

$$E_V(max) = \frac{Vel\,t_{RI}}{S_R} = R_\varepsilon(Vel)\,D \tag{3-124}$$

$$t_{RI} = R_{RI}\,C_{RI} = \frac{R_\varepsilon(Vel)\,D\,S_R}{Vel(max)} \tag{3-125}$$

Choose a reasonable value for C_{RI} and calculate R_{RI} using Equation (3-125).

C. Calculate R_1, R_2, and C_V (Figure 3-24) knowing the desired loop ω_n and ζ. The solutions for t_1 and t_2 are complicated by the fact that ω_n and ζ are functions of t_1. Solving Equation (3-112) for t_1

216

$$\iota_1 = \frac{DS_R}{\iota_{RI} \, \omega_n^{\;2}} \qquad (3\text{-}126)$$

Solving Equation (3-114) for ι_2 and substituting Equation (3-126) for ι_1,

$$\iota_2 = R_2 \, C_V = \frac{2\zeta}{\omega_n} - \frac{\iota_{RI}}{DS_R} \qquad (3\text{-}127)$$

choose a reasonable value for C_V and solve Equation (3-127) for R_2.

The value for R_1 may be found via Equation (3-113).

$$\iota_1 = (R_1 + R_2) \, C_V = \frac{DS_R}{\iota_{RI} \, \omega_n^{\;2}} \qquad (3\text{-}128)$$

$$R_1 = \frac{1}{C_V} \left[\frac{DS_R}{\iota_{RI} \, \omega_n^{\;2}} - \iota_2 \right] \qquad (3\text{-}129)$$

A loop will now be designed using the Type II specification with the following exceptions:

$R_e(max) = 50 \text{ ft} \, (\Delta \text{Vel} = 400 \text{ ft/sec}), \qquad R_e(\text{final}) = 10 \text{ ft} \, (\text{Vel} = 1250 \text{ ft/sec}).$

Acceleration error will be calculated (Equation (3-119)) and is not a design parameter.

A. Determine $T_R(max)$, S_V, S_R and PRF (same as Type II design)

$$S_V = 41 \times 10^3 \text{ V/sec}$$

$$S_R = 12,000 \text{ ft/sec}$$

$$PRF = 2.5 \text{ kHz}$$

B. Calculate R_{RI} and C_{RI} (Equation (3-125))

$$t_{RI} = R_{RI} C_{RI} = \frac{R_\varepsilon(\text{Vel}) D S_R}{\text{Vel(max)}} \qquad (3\text{-}130)$$

$$R_{RI} C_{RI} = \frac{10 (0.05)(12 \times 10^3)}{1250} \qquad (3\text{-}131)$$

$$R_{RI} C_{RI} = 4.8 \qquad (3\text{-}132)$$

Let $C_{RI} = 10 \, \mu F$

$$R_{RI} = \frac{4.8}{10 \times 10^{-6}} = 480 \, k\Omega \qquad (3\text{-}133)$$

C. Calculate R_1, R_2, and C_V

The final desired range error is small (10 feet), thus Figure 3-8 can reasonably be used for our damping factor of 0.707 and a maximum ΔVel error of 50 feet.

$$X \simeq 0.45 \qquad (3\text{-}134)$$

$$\omega_n = \frac{X \, \Delta \text{Vel}}{R_\varepsilon(\text{max})} = \frac{0.45 (400)}{50} \qquad (3\text{-}135)$$

or

$$\omega_n = 3.6 \qquad (3\text{-}136)$$

From Equation (3-127) (using $D = 0.05$ volt/ft)

$$\iota_2 = R_2 C_V = \frac{2\zeta}{\omega_n} - \frac{\iota_{RI}}{D S_R} \qquad (3\text{-}137)$$

or

$$\iota_2 = R_2 C_V = \frac{2(0.707)}{3.6} - \frac{4.8}{(0.05)(12 \times 10^3)} \qquad (3\text{-}138)$$

$$\iota_2 = R_2 C_V = 384.8 \times 10^{-3} \qquad (3\text{-}139)$$

Let $C_V = 10\ \mu F$

$$R_2 = \frac{384.8 \times 10^{-3}}{10 \times 10^{-6}} = 38.5\ k\Omega \qquad (3\text{-}140)$$

From Equation (3-127)

$$R_1 = \frac{1}{C_V} \left| \frac{D S_R}{\iota_{RI}\, \omega_n^2} - \iota_2 \right| \qquad (3\text{-}141)$$

$$R_1 = \frac{1}{10 \times 10^{-6}} \left| \frac{(0.05)(12 \times 10^3)}{(4.8)(3.6)^2} - 384.8 \times 10^{-3} \right| \qquad (3\text{-}142)$$

$$R_1 = 964\ k\Omega \qquad (3\text{-}143)$$

The rest of the design (range ramp generator, range discriminator, etc.) is the same as the Type II loop (Figure 3-19).

219

The loop can be verified the same way as a Type II loop (Figures 3-22 and 3-23) provided (Equation (3-118) with $A_V = 1$)

$$\frac{DS_R}{\iota_{RI}} \gg \frac{\omega_n}{\zeta} \qquad (3\text{-}144)$$

$$\frac{DS_R}{\iota_{RI}} = \frac{(0.05)(12 \times 10^3)}{4.8} = 125 \qquad (3\text{-}145)$$

and

$$\frac{\omega_n}{\zeta} = \frac{3.6}{0.707} = 5.1 \qquad (3\text{-}146)$$

Thus Equation (3-144) is well satisfied.

The range error as a function of time due to acceleration has been given as (Equation (3-120))

$$R_\varepsilon(t) = \frac{(\Delta Accel)(t)}{\dfrac{DS_R}{\iota_{RI}}} \qquad (3\text{-}147)$$

Solving Equation (3-147) for time, for a given maximum range error due to acceleration:

$$t = \frac{DS_R \, R_\varepsilon(max)}{\iota_{RI} \, \Delta Accel} \qquad (3\text{-}148)$$

Assuming a 50-foot maximum range error for an acceleration of 180 ft/sec²,

$$t = \frac{(0.05)(12 \times 10^3)(50)}{(4.8)(180)} = 34.7 \text{ sec} \qquad (3\text{-}149)$$

The loop noise bandwidth for a Type I loop may be approximated by the Type II loop Equation (3-43)

$$BW_n \simeq \frac{\omega_n}{2} \left(\zeta + \frac{1}{4\zeta} \right) \text{Hz} \tag{3-150}$$

or

$$BW_n = \frac{3.6}{2} \left(2\,(0.707) + \frac{1}{4\,(0.707)} \right) \tag{3-151}$$

$$BW_n = 3.18 \text{ Hz} \tag{3-152}$$

Figure 3-25 illustrates the Type I range-tracking loop. Note that an instrumentation amplifier has replaced the Type II range discriminator amplifier. Modern instrumentation amplifiers have input currents that are quite low and prevent excessive capacitor discharge; also, the discriminator gain, A_Δ, can be varied by a single resistor.

The two range tracking loops presented are inherently stable. Reference 1 presents an excellent discussion of loop stability from a phase-locked loop perspective.

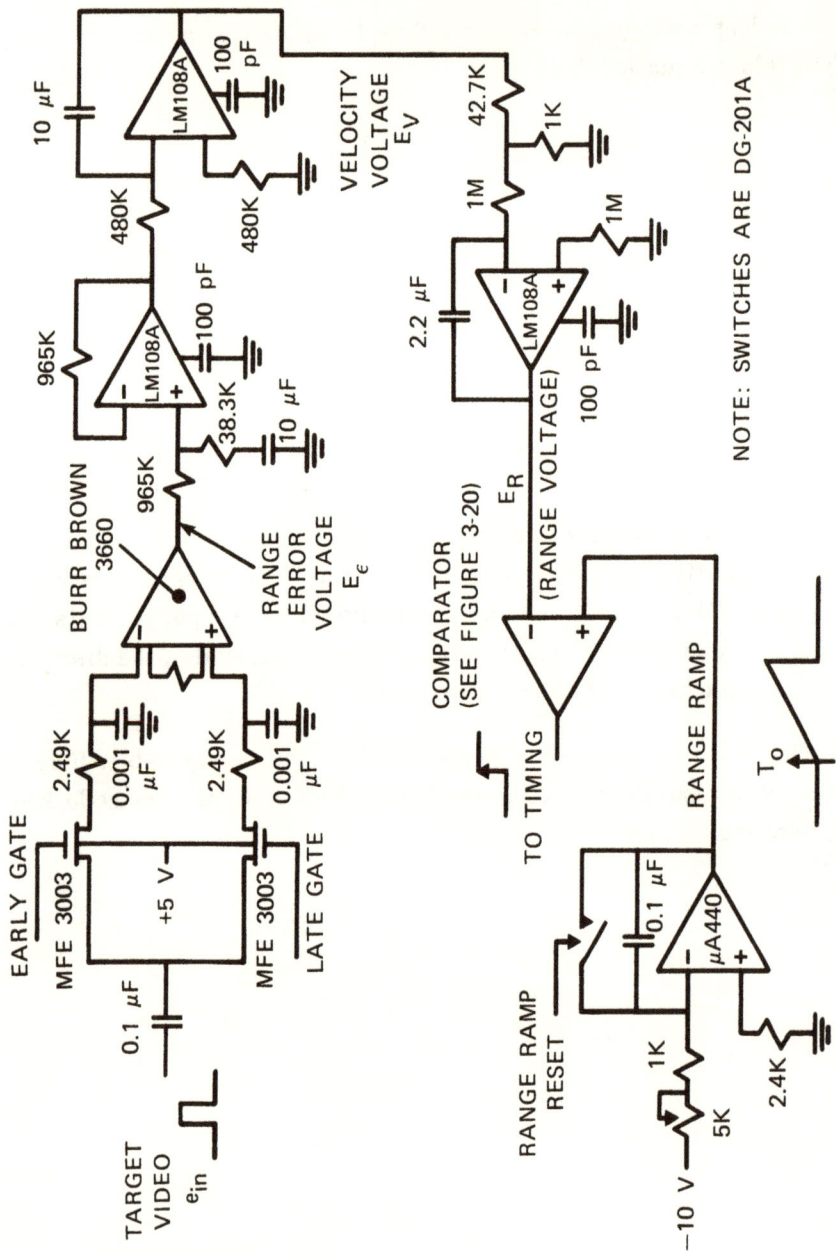

NOTE: SWITCHES ARE DG-201A

222

References

1. Blanchard, A. *Phase-Locked Loops: Application to Coherent Receiver Design*. John Wiley and Sons, New York, 1976.

2. Gardner, F. M. *Phaselock Techniques*. John Wiley and Sons, New York, 1966.

3. Hughes, R. S. *Logarithmic Amplification with Application to Radar and EW*. Artech House, Dedham, MA, 1986.

Appendix 3A

RANGE-TRACKING LOOP
AND PHASE-LOCKED LOOP ANALOGY

Figure 3A-1 illustrates the block diagram for a phase-locked loop. The basic operation of this loop is straightforward: the phase detector output is a function of the phase difference between the input, θ_{in}; and output, θ_{out}, and is given as

$$E_1 = K_1 (\theta_{in} - \theta_{out}) \qquad (3A\text{-}1)$$

where

K_1 = phase detector scale factor (volts/radian)

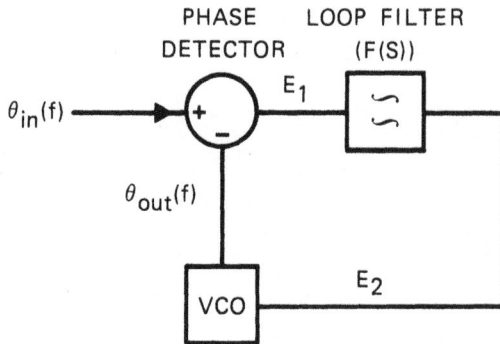

FIGURE 3A-1. Phase-Locked Loop.

The phase error is filtered (by $F(S)$) to suppress noise and determine dynamic loop performance.

The frequency of the voltage controlled oscillator (VCO) is controlled by the filter output voltage, E_2. The deviation of the VCO is

$$\Delta\omega = K_2 E_2 \qquad \text{(3A-2)}$$

where

K_2 = VCO sensitivity, Hz/volt = 1/2n (rad/sec/volt)

Since frequency is the derivative of phase, the VCO output may be given as

$$d\,\theta_{out}/dt = K_2 E_2 \text{ rad/sec} \qquad \text{(3A-3)}$$

and by taking the Laplace transform we obtain

$$L\left[\frac{d\,\theta_{out}(f)}{dt}\right] = S\,\theta_{out}(S) = K_2 E_2(S) \qquad \text{(3A-4)}$$

Thus, the phase output of the VCO is

$$\theta_{out}(S) = \frac{K_2 E_2(S)}{S} \qquad \text{(3A-5)}$$

or the phase of the VCO output is proportional to the integral of the control voltage, E_2. We will now relate the range tracking to that of a phase-locked loop.

Referring to Figures 3-3 and 3-4, at time, T_o, the T/R switch allows the transmitter signal to be radiated, and starts the range ramp. The range ramp slope, S_v, is

$$S_v = \frac{E_{R,max}}{T_{R,max}} \text{ V/sec} \qquad (3A\text{-}6)$$

where

$E_{R,max}$ = maximum ramp voltage

$T_{R,max}$ = maximum ramp width

The maximum ramp width must be at least as long as the maximum time for a signal return, or

$$T_{R,max} \geq \left(\text{Maximum range in feet} \right) \left(\frac{1 \text{ sec}}{492 \times 10^3 \text{ ft}} \right) \qquad (3A\text{-}7)$$

The range ramp slope may now be given as

$$S_v = \frac{(E_{R,max})(492 \times 10^3)}{R_{max}} \text{ V/ft} \qquad (3A\text{-}8)$$

Referring to Figures 3-3 and 3-4a, the range gate trigger occurs when the range ramp voltage equals the target range voltage. The target range voltage, E_R, is

$$E_R = \frac{E_V}{S R_{RI} C_{RI}} = \frac{E_V}{S \tau_{RI}} \qquad (3A\text{-}9)$$

where

E_V = range integrator input voltage (or velocity voltage)

ι_{RI} = integrator time constant, $R_{RI} C_{RI}$

The measured range voltage, R_o, may now be given as

$$R_o = \frac{E_R \text{ (volt)}}{S_V \text{ (volt/ft)}} \qquad \text{(3A-10)}$$

or substituting Equation (3A-10) into (3A-9),

$$R_o(S) = \frac{E_V \text{ (volt)}}{S \, \tau_{RI} \text{(sec)} \, S_V \text{ (volt/ft)}} \text{ ft/sec} \qquad \text{(3A-11)}$$

letting

$$S_R = \frac{1}{S_V} \text{ ft/V} \qquad \text{(3A-12)}$$

and

$$R_o(S) = \left(\frac{S_R}{\tau_{RI}}\right) \frac{E_V}{S} \qquad \text{(3A-13)}$$

where S_R/ι_{RI} for the range-tracking loop is the K_2 for the phase-locked loop (Equation (3A-2)).

Thus, the range integrator and range ramp may be replaced, for reasons of analysis, by Figure 3A-2.

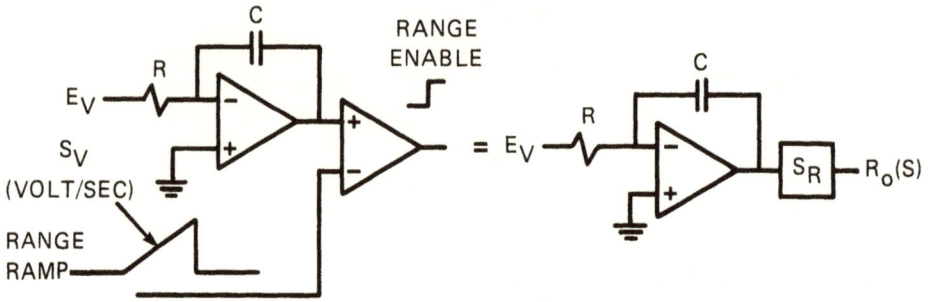

FIGURE 3A-2. Range Integrator Analysis Simplification.

Appendix 3B

RANGE DISCRIMINATOR ANALYSIS

Figure 3B-1 illustrates the basic early-late gate configuration. With the video centered as shown, the charge on each capacitor is the same and e_ε is zero. If more of the received signal is in the early gate, the difference amplifier will be positive (and vice-versa).

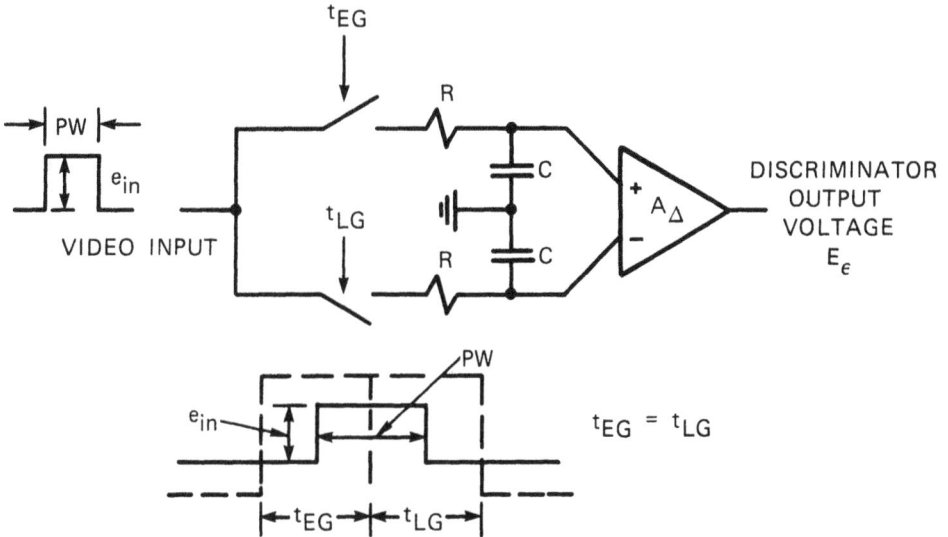

FIGURE 3B-1. Range Discriminator.

The range discriminator scale factor will be found with the aid of Figure 3B-2. The output voltage is a function of RC, PW, e_{in}, t, and the number of switch closures (effective closure time).

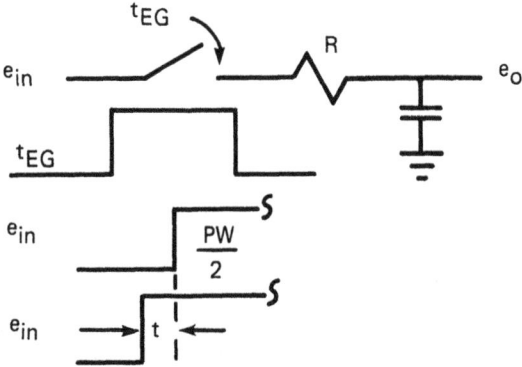

FIGURE 3B-2. Early Gate Timing.

The effective time constant may be given as

$$\iota_{eff} = \frac{PRI}{t_{EG}} (RC) \tag{3B-1}$$

where

$$PRI = \frac{1}{PRF} \tag{3B-2}$$

the output voltage (after 5 t_{eff} and assuming $RC \gg t_{EG}$)

$$e_o = \frac{PW/2}{t_{EG}} \, e_{in} \qquad \text{(3B-3)}$$

If the received pulse shifts into the early gate by t μsec, Figure 3B-2, the output voltage is

$$e_o(t) = \frac{PW/2 + t}{t_{EG}} \, e_{in} \qquad \text{(3B-4)}$$

Thus the net output of the range discriminator, Figure 3B-1 will be

$$E_\varepsilon(t) = A_D \left[\frac{PW/2 + t}{t_{EG}} - \frac{PW/2 - t}{t_{LG}} \right] e_{in} \qquad \text{(3B-5)}$$

and letting $t_{EG} = t_{LG}$,

$$E_\varepsilon(t) = A_D \left[\frac{2t}{t_{EG}} \right] e_{in} \qquad \text{(3B-6)}$$

The range discriminator scale factor may be found by differentiating Equation (3B-5) with respect to t

$$\frac{d\,E_\varepsilon(t)}{dt} = \frac{2\,A_D\,e_{in}}{t_{EG}} \quad \text{(V/sec)} \qquad \text{(3B-7)}$$

Multiplying Equation (3B-6) by 1 sec/492 $\times$ 10^6 ft, the range discriminator slope is obtained

$$D = \frac{2\,A_D\,e_{in}}{t_{EG}\,(492 \times 10^6)} \quad \text{(V/ft)} \qquad \text{(3B-8)}$$

or if t_{EG} is in nanoseconds,

$$D = \frac{2 A_D e_{in}}{(0.492) t_{EG} (nsec)} \quad (V/ft) \qquad (3B\text{-}9)$$

or

$$D \simeq \frac{4.1 A_D e_{in}}{t_{EG} (nsec)} \quad (V/ft) \qquad (3B\text{-}10)$$

The value for RC must be much greater than t_{EG}, and $5 \tau_{eff}$ must be much less than the loop closure times (t_c). Assume the following loop conditions:

$$t_{EG} = 100 \text{ nsec}$$

$$e_{in} = 3 \text{ volts}$$

$$RF = 2 \text{ kHz (or PRF} = 50 \text{ μsec)}$$

$$\text{loop closure time} = 0.3 \text{ second}$$

Letting

$$RC \geq 10 t_{EG} \qquad (3B\text{-}11)$$

or

$$RC \geq 1 \text{ μsec} \qquad (3B\text{-}12)$$

Thus,

$$RC \geq 1 \times 10^{-6} \qquad (3B\text{-}13)$$

Let $R = 1 \text{ k}\Omega$

$$C \geq 0.001 \text{ μF} \qquad (3B\text{-}14)$$

232

Letting

$$5\,t_{eff} \leq \frac{t_c}{10} \tag{3B-15}$$

$$5\left(\frac{RC}{t_{EG}\,PRF}\right) \leq \frac{t_o}{10} \tag{3B-16}$$

$$RC \leq \frac{t_o\,t_{EG}\,PRF}{50} \tag{3B-17}$$

or

$$RC \leq 0.02\,t_o\,t_{EG}\,PRF \tag{3B-18}$$

and

$$(1\,K)(0.001 \times 10^{-6}) \leq (0.02)(0.3)(100 \times 10^{-9})(2 \times 10^3) \tag{3B-19}$$

or

$$1 \times 10^{-6} \leq 4 \times 10^6 \tag{3B-20}$$

and the necessary range discriminator conditions are met.

Figure 3B-3 summarizes the range discriminator equations.

Figures 3B-4, 3B-5, and 3B-6 illustrate the general shape of the range discriminator output for several PW/t_{EG} conditions ($t_{EG} = t_{LG} = t$).

233

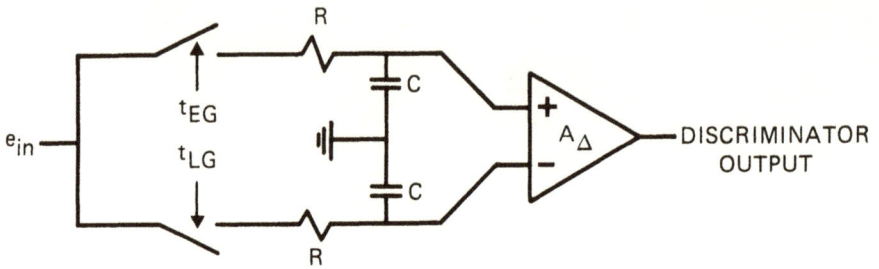

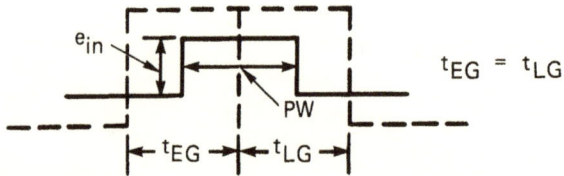

$$RC \geqslant 10\ t_{EG} \qquad\qquad RC \leqslant 0.02\ t_o\ t_{EG}\ PRF$$

$$t_o = \text{RANGE LOOP CLOSURE TIME (SEC)}$$
$$t_{EG} = \text{EARLY GATE WIDTH (SEC)}$$

$$D = \frac{4.1\ A_\Delta\ e_{in}}{t_{EG}\ (nsec)} \quad \frac{VOLT}{FT}$$

FIGURE 3B-3. Range Discriminator Design Summary.

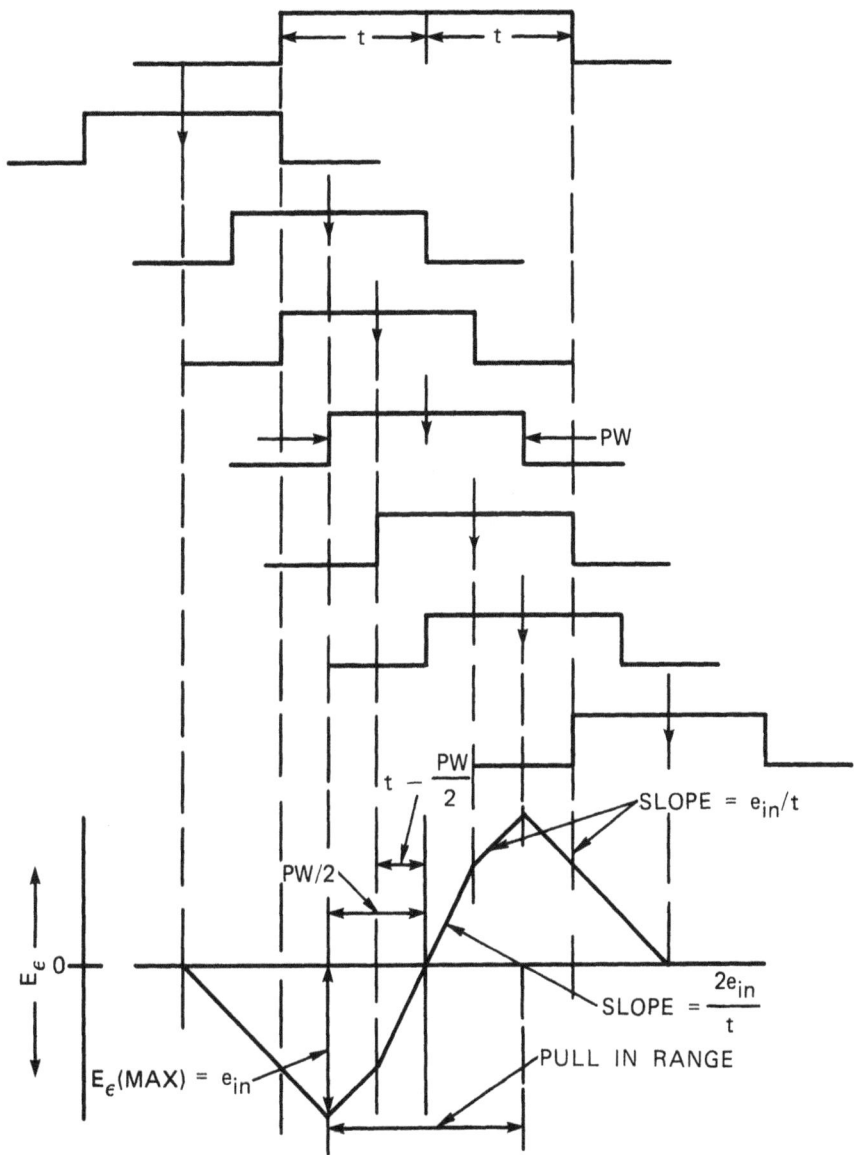

FIGURE 3B-4. Range Discriminator Output for PW > t.

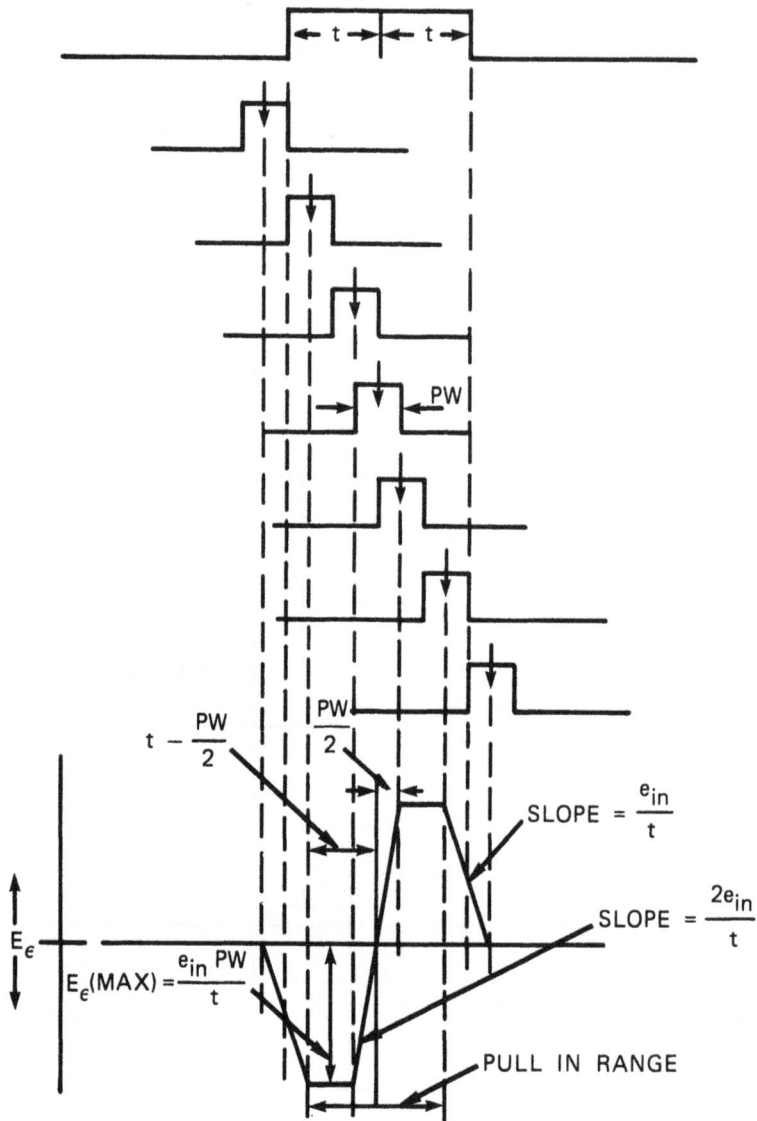

FIGURE 3B-5. Range Discriminator Output for PW < t.

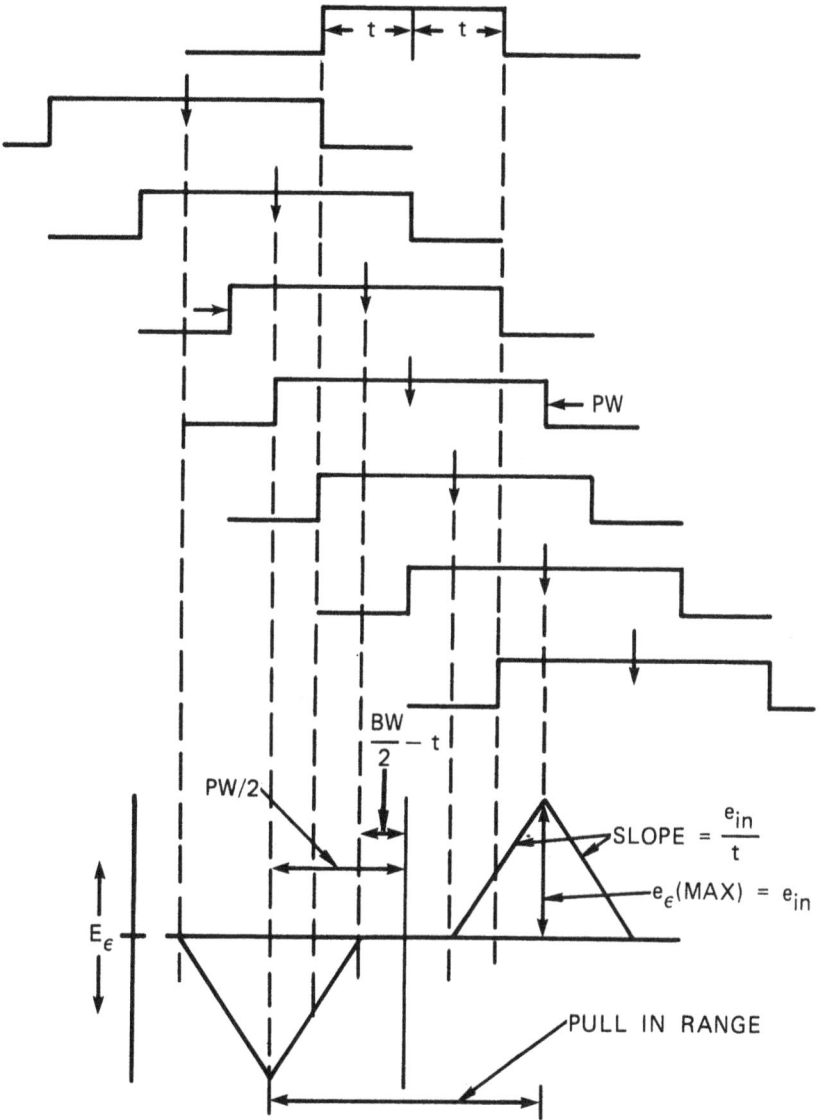

FIGURE 3B-6. Discriminator Output for PW > 2t.

Nomenclature

A_Δ differencing amplifier gain

BW_n loop noise bandwidth

c speed of light (it is obvious when not meant as capacitance)

D range discriminator scale factor (volt/ft)

E_ε range error voltage
E_R target range voltage
$E_s(max)$ maximum range ramp voltage
E_V target velocity voltage
E_{SM} maximum range ramp sweep voltage
e_ε range discriminator output voltage
e_{in} target video return

$F(S)$ filter transfer function

$H(S)$ loop transfer function

K dc loop gain (D_{SR}/τ_{RI})
K_1 phase-locked loop phase detector scale factor (volt/radian)
K_2 phase-locked loop VCO sensitivity (Hz/volt)

LG loop gain

238

PRF pulse repetition frequency
PRI pulse repetition interval
PW pulse width

R range (it is obvious when not meant as resistance)
R_ε range error
R_{in} effective input range
$R_{max}(ft)$ maximum range in feet
R_o effective output range

S Laplacian S
SE_ε discriminator scale factor
S_R range integrator scale factor (ft/volt)
S_v range ramp scale factor (volts/sec)

T_0 master trigger time (range ramp start)
T_R target range time
T_{RS} track loop reset time
T_{SW} range ramp sweep width
t Early gate and late gate widths
t_c loop closure time
t_{EG} early gate width
t_{LG} late gate width

Δt time change
ΔR range step
ΔVel velocity step
$\Delta Accel$ acceleration step
ζ loop damping factor
τ_1, τ_2 velocity integrator time constants
τ_{RI} range integrator time constant
θ phase-locked loop phase
ω_n loop natural frequency

Bibliography

Lock, A. *Guidance.* D. Van Nostrand Company, Inc., 1955, pp. 408-415.

Drogin, E. "Building a Range Tracker with Digital Circuits." *Electronics,* 27 March 1972, pp. 86-90.

Naval Weapons Center. *A Practical Approach to the Analysis and Design of Zero Velocity Error Range Trackers,* by Richard Smith Hughes. China Lake, CA, NWC, December 1973. (NWC TP 5566, publication UNCLASSIFIED.)

INDEX

Automatic gain control
  (AGC), 16,45,64,96
Automatic noise tracking, 111

Bandwidth, 14

Calculations
    AGC gain, 96
    loop rise time, 99
    static regulation, 93

Conical scanning, 64

Continuous wave (CW), 1

Design
    equations, 215
    example, 195
    procedure, 185
    verification, 16
Detector characteristics, 84
Dynamic regulation, 3,8

Electronic warfare (EW), 1

FETs, GaAs, 55

GaAs (Gallium Arsenide), 55

Input modular reduction (IMR), 3
Integrated circuits (IC), 55
Intermediate frequency (IF), 1

Linear detector, 28
Linearized time constant, 75
Loop gain (LG), 3
Loop rise time, 3,14,74,99
Loop stability, 80
Loops
    automatic noise tracking, 111
    conical scanning AGC, 64
    phase-locked, 224
    range tracking, 175,213,224
Low-pass filter (LPF), 1

Modular reduction, input (IMR), 3
Monolithic microwave ICs
  (MMIC), 55

Noise, 135
Nomenclature, 104,171,238

Phase-locked loop, 224
PIN diodes, 49
P-N junction diodes, 40,45
Power-voltage relationships, 83

Radio frequency (RF), 1
   gain, 138
Range discriminator, 229
Range tracker design, 185
Range tracking loop, 213,224
Receiver gain, 135

Schottky diode, 46,68
Signal sensitivity, 135,154
Square law detector, 16

Static regulation, 3,4,93
Symbols, *see* Nomenclature

Tangential sensitivity, 138
Test circuit, 16,28
Thresholding, 150
Time constant, 75

Variable gain elements, 36,40,49,55
Variable time constant, 75

www.ingramcontent.com/pod-product-compliance
Lightning Source LLC
Chambersburg PA
CBHW021430180326
41458CB00001B/200